Tiansi Dong

Recognizing Variable Environments

Studies in Computational Intelligence, Volume 388

Editor-in-Chief
Prof. Janusz Kacprzyk
Systems Research Institute
Polish Academy of Sciences
ul. Newelska 6
01-447 Warsaw
Poland
E-mail: kacprzyk@ibspan.waw.pl

Further volumes of this series can be found on our homepage: springer.com

Vol. 365. Roger Lee (Ed.)
Computers, Networks, Systems, and Industrial Engineering 2011, 2011
ISBN 978-3-642-21374-8

Vol. 366. Mario Köppen, Gerald Schaefer, and Ajith Abraham (Eds.)
Intelligent Computational Optimization in Engineering, 2011
ISBN 978-3-642-21704-3

Vol. 367. Gabriel Luque and Enrique Alba
Parallel Genetic Algorithms, 2011
ISBN 978-3-642-22083-8

Vol. 368. Roger Lee (Ed.)
Software Engineering, Artificial Intelligence, Networking and Parallel/Distributed Computing 2011, 2011
ISBN 978-3-642-22287-0

Vol. 369. Dominik Ryżko, Piotr Gawrysiak, Henryk Rybinski, and Marzena Kryszkiewicz (Eds.)
Emerging Intelligent Technologies in Industry, 2011
ISBN 978-3-642-22731-8

Vol. 370. Alexander Mehler, Kai-Uwe Kühnberger, Henning Lobin, Harald Lüngen, Angelika Storrer, and Andreas Witt (Eds.)
Modeling, Learning, and Processing of Text Technological Data Structures, 2011
ISBN 978-3-642-22612-0

Vol. 371. Leonid Perlovsky, Ross Deming, and Roman Ilin (Eds.)
Emotional Cognitive Neural Algorithms with Engineering Applications, 2011
ISBN 978-3-642-22829-2

Vol. 372. António E. Ruano and Annamária R. Várkonyi-Kóczy (Eds.)
New Advances in Intelligent Signal Processing, 2011
ISBN 978-3-642-11738-1

Vol. 373. Oleg Okun, Giorgio Valentini, and Matteo Re (Eds.)
Ensembles in Machine Learning Applications, 2011
ISBN 978-3-642-22909-1

Vol. 374. Dimitri Plemenos and Georgios Miaoulis (Eds.)
Intelligent Computer Graphics 2011, 2011
ISBN 978-3-642-22906-0

Vol. 375. Marenglen Biba and Fatos Xhafa (Eds.)
Learning Structure and Schemas from Documents, 2011
ISBN 978-3-642-22912-1

Vol. 376. Toyohide Watanabe and Lakhmi C. Jain (Eds.)
Innovations in Intelligent Machines - 2, 2011
ISBN 978-3-642-23189-6

Vol. 377. Roger Lee (Ed.)
Software Engineering Research, Management and Applications 2011, 2011
ISBN 978-3-642-23201-5

Vol. 378. János Fodor, Ryszard Klempous, and Carmen Paz Suárez Araujo (Eds.)
Recent Advances in Intelligent Engineering Systems, 2011
ISBN 978-3-642-23228-2

Vol. 379. Ferrante Neri, Carlos Cotta, and Pablo Moscato (Eds.)
Handbook of Memetic Algorithms, 2011
ISBN 978-3-642-23246-6

Vol. 380. Anthony Brabazon, Michael O'Neill, and Dietmar Maringer (Eds.)
Natural Computing in Computational Finance, 2011
ISBN 978-3-642-23335-7

Vol. 381. Radosław Katarzyniak, Tzu-Fu Chiu, Chao-Fu Hong, and Ngoc Thanh Nguyen (Eds.)
Semantic Methods for Knowledge Management and Communication, 2011
ISBN 978-3-642-23417-0

Vol. 382. F.M.T. Brazier, Kees Nieuwenhuis, Gregor Pavlin, Martijn Warnier, and Costin Badica (Eds.)
Intelligent Distributed Computing V, 2011
ISBN 978-3-642-24012-6

Vol. 383. Takayuki Ito, Minjie Zhang, Valentin Robu, Shaheen Fatima, and Tokuro Matsuo (Eds.)
New Trends in Agent-based Complex Automated Negotiations, 2011
ISBN 978-3-642-24695-1

Vol. 384. Daphna Weinshall, Jörn Anemüller, and Luc van Gool (Eds.)
Detection and Identification of Rare Audiovisual Cues, 2011
ISBN 978-3-642-24033-1

Vol. 385. Alex Graves
Supervised Sequence Labelling with Recurrent Neural Networks, 2012
ISBN 978-3-642-24796-5

Vol. 386. Marek R. Ogiela and Lakhmi C. Jain (Eds.)
Computational Intelligence Paradigms in Advanced Pattern Classification, 2012
ISBN 978-3-642-24048-5

Vol. 387. David Alejandro Pelta, Natalio Krasnogor, Dan Dumitrescu, Camelia Chira, and Rodica Lung (Eds.)
Nature Inspired Cooperative Strategies for Optimization (NICSO 2011), 2012
ISBN 978-3-642-24093-5

Vol. 388. Tiansi Dong
Recognizing Variable Environments, 2012
ISBN 978-3-642-24057-7

Tiansi Dong

Recognizing Variable Environments

The Theory of Cognitive Prism

Author

Dr. Tiansi Dong
AG Prof. Helbig
Informatikzentrum
Universitätsstr. 1
58084 Hagen
Germany
E-mail: tiansi.dong@fernuni-hagen.de

ISBN 978-3-642-24057-7 e-ISBN 978-3-642-24058-4

DOI 10.1007/978-3-642-24058-4

Studies in Computational Intelligence ISSN 1860-949X

Library of Congress Control Number: 2011938289

Typeset & Cover Design: Scientific Publishing Services Pvt. Ltd., Chennai, India.

Printed on acid-free paper

9 8 7 6 5 4 3 2 1

springer.com

To my parents, Peiling, and Sophia

Foreword

Tiansi Dong sent me an e-mail asking whether he can read for some days a book which I borrowed from the library. This way I came in contact with a young scholar sitting not more than 150 m away in another building on the same campus. I went to his office to bring him the book, and interested in his research work, in particular, his motivation to read the book. Our chatting soon went beyond the book. He told me about his background, how he came to Germany and what brought him to Fernuniversität in Hagen. Especially interesting was what he experienced as doctoral student in Bremen, where his research was integrated in the framework of the so-called "Cluster of Excellence for Spatial Cognition".

Dong's doctoral work challenged several basic assumptions of qualitative spatial representation, established an axiom governing the connection relation which is missing in the well-known Region Connection Calculus and neglected in the formalism of Whitehead's *Process and Reality*. Dong's work does not only lead to the fact that all qualitative orientation frameworks can be unified, but also that qualitative and quantitative orientation relations form a continuum. Having made these remarkable contributions he spent a couple of years in understanding why his supervisor discouraged him to submit his results to journals for publication and, even more, that a "Cluster of Excellence" shall not just praise itself as excellent, rather shall justify its existence by making contributions to the tax-payer.

After his doctorate, with great efforts Dong managed to publish some of his results in scientific journals and at conferences. Also, he received some recognition by being invited to serve on the panel of Zentralblatt MATH, maintained by the European Mathematical Society, Leibniz Institute for Information Infrastructure Karlsruhe, Heidelberg Academy of Sciences and Humanities, and Springer-Verlag. Nevertheless, due to many obstacles Dong did not succeed in publishing his results in their entirety. Therefore, I persuaded him to choose a way where he would not encounter censorship in form of reviewing, viz. as a book, which, I convinced him, is the genuine form of publication – books are read even after centuries, whereas journal contributions are considered obsolete within just a few years.

The resulting book you hold in your hands starts with an interesting story to motivate its arguments in a natural way, analyzing the problem considered and laying ground in psychology and philosophy for the results developed later. Readers can enjoy Dong's interesting ideas and solid arguments from Chapter 1 through 4 without any knowledge of mathematics. To understand Chapter 5, readers are assumed to know a little bit about first-order logic. Chapter 6 should be interesting to programmers and engineers who like to implement algorithms.

Hagen, May 2011 Wolfgang A. Halang

Preface

This book is based on my research that has spanned over a couple of last years, resulted in my PhD dissertation and papers after that. To put it briefly, the topic of my research, including that of my PhD dissertation, is how dynamic environments could be recognized by artificial systems. The focus was placed on identifying a simple method to represent spatial knowledge, which anchors as much as possible to cognitive psychology, linguistics, philosophy and neurology. One of the results obtained is an axiom system which represents spatial knowledge. If this system is correct, the Region Connection Calculus (RCC), currently prevailing in science for the purpose of spatial knowledge representation, will turn out to be incomplete and flawed, and all orientation frameworks could be considered as trivial cases of the system presented here. This is indeed the case.

This book should not only be interesting for researchers in the field of spatial knowledge representation and reasoning, but also for scientists working in areas related to computational intelligence, for which spatial intelligence plays a fundamental role. Since the book demonstrates a way to integrate research results from different fields into a unified system, it should also be interesting for those carrying out interdisciplinary research. Particularly, I dedicate this book to talented doctoral students to make them aware of the possibility that ground-breaking research results they may come up with are not welcomed, and that they may even encounter difficulties in their lives. Allan Baddeley mentioned this in his lecture during the Summer School in Cognitive Science'03 in Sofia, Bulgaria, which I tested in my PhD research and proven to be correct.

I am indebted to the Chinese Christian communities in Germany for their understanding and support during my difficult times – if there is an original contribution in this book, it will be to *give glory to our Father in heaven*. I am definitely indebted to Wolfgang Halang, Janusz Kacprzyk and Thomas Ditzinger who helped me publishing this book in high quality in a renowned book series for an easy accessibility to the scientific community.

Hagen, May 2011 Tiansi Dong

Contents

List of Figures

List of Tables

Abstract

Objects are not statically located in environments. Therefore, a snapshot view of an environment at one moment may be quite different from its snapshot at other moments. However, people are capable of recognizing the environment. This thesis presents a theory of recognizing variable spatial environments — The Theory of Cognitive Prism. It collected theoretical and empirical evidences from several disciplines, such as cognitive psychology, neurology, psycho-linguistics, philosophy, and proposes a symbolic computational theory. This thesis has three parts: The first part is the description of the theory in natural language. It presents a commonsense knowledge of the snapshot view of a spatial environment—the "cognitive spectrum", and of the process of recognizing a spatial environment by comparing two cognitive spectrums. The second part is the mereotopological formalism of the theory. It presents the formal structure of the cognitive spectrum and the process of comparing two formalized cognitive spectrums. The third part introduces a symbolic simulating system of the theory — the LIVE model.

The research questions are: (1) What are the spatial relations between extended objects through observation? (2) According to what principle is the reference ordering between extended objects formed? (3) What is the structure of a cognitive spectrum? (4) How shall two cognitive spectrums be compared? (5) How does the difference between two cognitive spectrums determine the degree of the compatibility between the perceived cognitive spectrum and the target one? (6) How shall the results of Question (1) to (5) be formalized?

Distance relations between extended objects are understood by the degree of the extension from one extended object to the other; orientation relations between two extended objects are distance comparison between one extended object and sides of the other extended object. The reference ordering between two extended objects is based on the commonsense knowledge of relative stabilities of related objects. The cognitive spectrum is structured by extended objects and their spatial relations which obey the reference ordering. Two cognitive spectrums are compared based on the categories and locations of extended objects. The compatibility of two cognitive spectrums is determined by their differences and relative stabilities of related objects. Recognizing spatial environments is the comparison process between the cognitive spectrum of the perceived environment and the cognitive spectrum of target environment and the judgement of the compatibility between them. All above are mereotopolgically formalized by the connectedness relationship **C**. The computational complexity of the formalized recognition process is *polynomial* P. A symbolic simulation system, the LIVE model, has been implemented in Lisp.

Chapter 1
An Introduction

Mr. Bertel has lately rented an unfurnished single-room apartment as his home, shown in Figure 1.1(a). The door of the apartment is in the middle of one wall. There is a big window opposite to the door.

On the next day Mr. Bertel buys a writing-desk, a bookshelf, and a big couch, Figure 1.1(b). He puts the writing-desk next to the window for good eyesight. He puts the bookshelf to the left side of the writing-desk. The couch that is used as bed at night is put left to the door.

On the third day Mr. Bertel buys a balloon as a chair, which is good for his back, a small tea-table, and a dining-table, Figure 1.1(c). The balloon is put in front of the writing-desk; the small tea-table is put in front of the couch; the dining-table is close to the wall opposite to the couch.

On the fourth day Mr. Bertel buys some flowers, a lamp, several tea-cups, some books, and a picture, Figure 1.1(d). He puts the lamp on the writing-desk, the flowers on the table. The picture is hung on the wall to decorate his new home. The books are put on the bookshelf. The tea-cups are placed on the tea-table.

On the fifth day Mr. Bertel's mother comes to see her son's new beautiful home.

On the sixth day Mr. Bertel invites his neighbor, Mr. Certel, to his new home. Mr. Certel gives Mr. Bertel a small plastic-bound red book[1], which is put on the writing-desk, and a bunch of flowers that are put on the dining-table. Mr. Certel likes sitting on the balloon to chat with Mr. Bertel who is lying on the couch, Figure 1.1(e).

On the next day Mr. Bertel's mother comes and mumbles, why did her son put the balloon to the tea-table? She moves the balloon back to the writing-desk without paying attention to the new book and the flowers from Mr. Certel.

On the eighth day the flowers are withered and thrown away. All the books are put on the bookshelf. In the evening Mr. Bertel moves his writing-desk to the right side of the bookshelf to make room for a party. The balloon is put near the table which is now at the right hand side of the door in the corner, Figure 1.1 (f).

On the ninth day Mr. Bertel's mother comes again. This time she goes into Mr. Certel's apartment by mistake, Figure 1.1(g), while Mr. Certel joined Mr. Bertel's

[1] Mao Tse Tung's Bible.

T. Dong: Recognizing Variable Environments, SCI 388, pp. 3–10.
springerlink.com

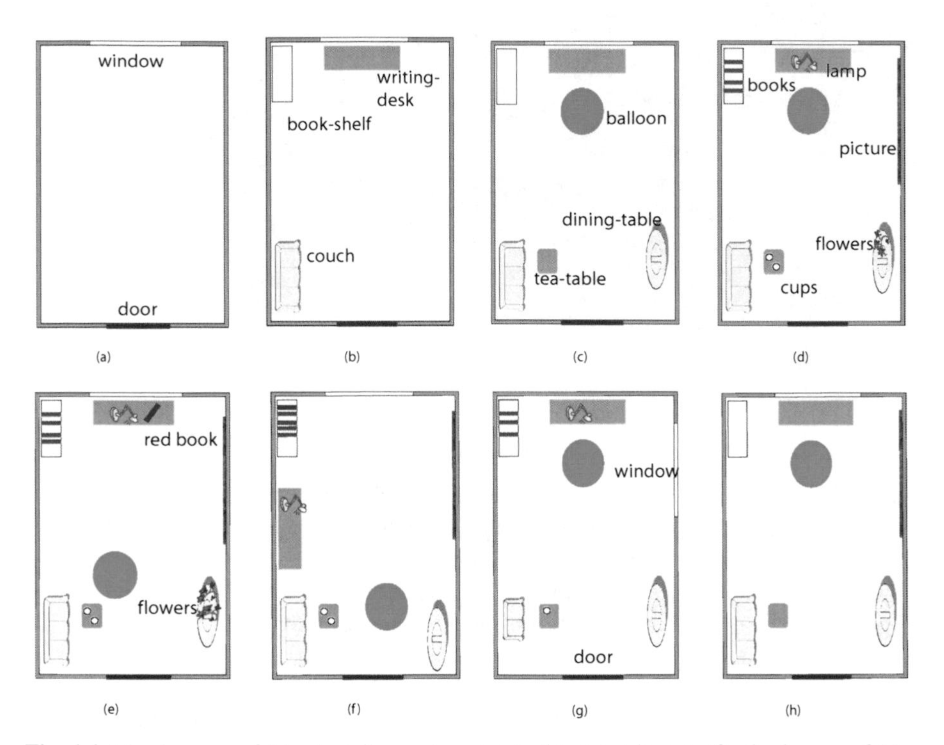

Fig. 1.1 The layouts of Mr. Bertel's apartment (a), (b), (c), (d), (e), (f); the layout of Mr. Certel's apartment (g); the layout of Mr. Bertel's apartment as his mother remembers (h)

party without closing his door, got tired, and fell asleep on Mr. Bertel's dining-table. Although Mr. Certel's apartment is very similar to Mr. Bertel's apartment, Mr. Bertel's mother notices that she is not in her son's home and leaves the apartment. When she goes to Mr. Bertel's apartment, she finds that the writing-desk is located between the bookshelf and the couch and that the table and the balloon are located differently as she expects, she wonders for a while and accepts it as her son's apartment. When she meets Mr. Bertel, she asks what happened last night.

How is Mr. Bertel's mother able to recognize that she is not in her son's apartment? The window in the perceived environment is located differently and the couch is a little shorter. How could her son have changed this? When Mr. Bertel's mother gets to Mr. Bertel's apartment, he is in the kitchen and Mr. Certel is in the bathroom. However, she is confident that she is in the right apartment, because if the writing-desk along with the balloon is put next to the window, and the table is moved closer to the picture, the apartment looks just like what she has seen before, Figure 1.1(h). She does not mind the absence of the flowers or the different number of tea-cups — one of the tea-cups was broken during the party.

1.1 The Aim

The layout of Mr. Bertel's apartment changes to meet his needs, such as to chat with a friend or to hold a party. The configurations of our spatial environments often change, e.g., our offices, kitchens, dining-rooms, sitting-rooms. In our offices chairs are moved very often; books and stationeries are sometimes put on the desk and sometimes into drawers. At our homes plates and bowls are on the table during eating time, after that they are put in the sink and then in the closet. We can see all these changes; however, we do not suspect that they are not our offices or our homes just because of the changed layouts. The aim of this thesis is on the commonsense knowledge needed for recognizing variable spatial environments and to make computational modelling. It researches into the question of how to recognize a remembered spatial environment with regards to the nature of its variable layout.

1.2 How Can We Recognize Variable Spatial Environments?

How can Mr. Bertel's mother distinguish Mr. Bertel's apartment from Mr. Certel's apartment? The answer appears simple: Because Mr. Bertel's apartment is different from Mr. Certel's. However, if we notice that Mr. Bertel's apartment is not a static environment, we understand the difficulty in the question: How does Mr. Bertel's mother distinguish her son's apartment from other rooms with regards to the changing layouts of all these rooms? Mr. Bertel's mother must have some knowledge of her son's apartment. What kind of knowledge does she have? How can she use the knowledge of Mr. Bertel's apartment to justify whether the perceived apartment is her son's apartment or not?

Some people say that they know which is their office since their office key can open the door; some people say that they know which is their office, because their puppy bear sits on the desk in the office; some say that they know which is their office, because their office door is the biggest on the floor; etc.

What these people say reflects the simplest situation to recognize a spatial environment. They recognize a representative figural object in the spatial environment, with which they can recognize the spatial environment. For example, when we see the Eiffel Tower, we know that we are in Paris; when we see Big Ben, we know that we are in London; when we see the Great Wall, we know that we are in Peking; when we see the Statue of Liberty, we know that we are in New York. However, what happens when the key can open many rooms? What happens when the puppy bear is in a drawer? etc.

This simplest method also fails for cognitive agents who cannot recognize such a representative figural object. For example, a robot has just visited your office and on your order it moves to the kitchen to fetch a cup of coffee. Consider the simplest situation: There is only a corridor connecting your office and the kitchen. The robot must go out of your office, pass the corridor, and enter the kitchen; and then take the same way back. During this time your chair is moved a bit; there are people or other robots walking on the corridor; and the coffee machine in the kitchen may have been moved. With all these variabilities in spatial environments current robots may take much more time, if they can achieve this, than your going to the kitchen and making the coffee yourself.

A better way to recognize spatial environments may be to consider the environment as a whole, and to compare the spatial layout at two different times. When people get to a spatial environment, they have a snapshot view of the configuration of the environment. They compare the current perceived configuration with the one they remember, and make judgement of the transformation between them. Recognizing an environment can be interpreted as a kind of subjective judgment of the ease of transformation from the remembered configuration into the perceived configuration. If the transformation is very difficult, e.g., the perceived window is located at a different location, then a negative judgment will be made; if the transformation is easy, for example, differently located chairs and books, then a positive judgment will be made.

The following questions will, therefore, be addressed: How shall we represent a spatial configuration? How shall we represent a transformation from one configuration into another? And how shall we judge its ease?

1.3 Interdisciplinary Perspectives

The representation of spatial configurations can be explored from different perspectives.

1.3.1 Neurology

Neurology provides cases of brain-impaired patients who lost some of their mental abilities. By examining what mental abilities they lost and what they can still achieve, we can find causal relations among mental abilities. For example, Wilson et al. (1999) reported a patient who could not retrieve images from her long-term memory. The patient lost the ability to retrieve images, thus she cannot recognize single objects as precisely as normal people. On the other hand, she can still recognize spatial environments, like her home. This could be roughly understood by the findings in cognitive neuroscience that the visual system consists of at least two subsystems: the "what" cortical system and the "where" cortical system, e.g., Ungerleider and Mishkin (1982), Rueckl et al. (1989), Creem and Proffitt (2001). The patient's "what" cortical system is partially damaged, but her "where" cortical system is normal. Recognizing spatial environments relies more on the "where" cortical system than on the "what" cortical system. This case shows that **recognizing spatial environments does not require much information on figural objects** and that **spatial location of objects are important for recognizing spatial environments**.

1.3.2 Cognitive Psychology

Cognitive psychology explores human or animal behavior by performing empirical experiments. Tolman (1948) conducted a series of experiments on rats and showed that rats have "cognitive maps" about spatial environments.

Research of object recognition shows that humans identify objects[2] at certain level of categories, e.g., "basic level category" by Rosch et al. (1976), or "entry point level" by Jolicoeur et al. (1984). At the "basic level" humans are fastest to categorize instances, Rosch (1975), fastest to identify, Murphy and Smith (1982), and to spontaneously choose a name. The "basic level" object recognition shows that even normal people do not see everything of a single object, though they may see a bit more than the patient reported by Wilson et al. (1999). Thus, the results of neurology and the results of cognitive psychology have the convergent conclusion that **to recognize spatial environments requires to categorize objects, rather than to identify them**.

Psychological experiments on spatial relations found that humans use three kinds of spatial relations: topological relations, qualitative orientation relations, and qualitative distance relations, Piaget and Inhelder (1948).

Psychological research on mental spatial models showed that **spatial models are hierarchically structured and have distortions**, e.g., Tversky (1981), McNamara (1986), McNamara (1991), Tversky (1991), McNamara (1992), Tversky (1992).

1.3.3 Psycho-linguistics

Psycho-linguistics also provides a window to human cognition. The corresponding relationship between mental model of spatial environments and their language descriptions were discussed in Talmy (1983) and Tversky and Lee (1999). This provides theoretical support to explore a person's mental representation of spatial environments by examining her/his language descriptions of spatial layouts. The non-symmetric relation between the location object and the reference object in spatial linguistic descriptions supports the existence of relative stability. This thesis proposes that **people have commonsense knowledge of relative stabilities of objects in a spatial environment** and that **objects are referenced to more stable objects in an environment which leads to the hierarchical structure of the spatial configuration in the mind**.

1.3.4 Formal Spatial Ontologies

Grenon and Smith (2004) proposed that a dynamic spatial ontology should combine two distinct types of inventory of the entities and relationships in reality: On the one hand, a purely spatial ontology supporting snapshot views of the space at successive instants of time: SNAP; on the other hand, a purely spatiotemporal ontology of change and process: SPAN. SPAN is a 4-dimensionalist ontology of processes

[2] Except the recognition of human faces.

— it has temporal parts and unfolds itself phase by phase. Following this perspective, recognizing a spatial environment relates two snapshots (one is remembered in mind, the other is currently perceived) and the spatiotemporal relation between them: Can they (or to what extent can they) participate into one SPAN?

Intuitively, we do not recognize a place by checking everything in it. The air in your office, a sheet of paper, apples on a table and contents of the dustbin do not help recognize your office. This thesis raises the ontological question for the task of recognizing spatial environments: **What exists in an environment that makes it to be that environment?**

Smith and Varzi (2000) distinguish two basic typologies of boundaries: *Bona fide* (or physical) boundary and *fiat* (in the sense of human decision or delineation) boundary. Two basic typologies of spatial objects follow from this distinction: *Bona fide* objects which have *bona fide* boundaries and *fiat* objects which only have *fiat* boundaries. This thesis distinguishes *fiat* ontologies from *bona fide* ontologies following Smith and Varzi (2000), Smith (2001). It proposes that **recognizing spatial environments relates to *fiat* ontologies through perception, cognition, and language expressions**. It has little to do with finger-print or foot-print checking, molecule analysis, or DNA testing.

1.3.5 Computational Modelling

Computer science, especially artificial intelligence, provides methods of knowledge representation, symbolic system construction, and implementation to simulate cognitive phenomena. The philosophical assumption behind experimental computational modelling is that cognitive phenomena are computable, Newell and Simon (1972), Newell (1980). The method of computational modelling has been successfully applied to cognitive structures and processes of environmental spatial knowledge, e.g., Kuipers (1977), Kuipers (1978), Yeap (1988), Yeap and Jefferies (1999), Klippel (2003a), Klippel (2003b), Klippel et al. (2004), Klippel and Richter (2004), Klippel and Montello (2004), Klippel et al. (2005a), Klippel et al. (2005b), Klippel et al. (2005c) and of geographic spatial knowledge, e.g., Barkowsky (2001), Barkowsky (2002), Barkowsky (2003), Engel et al. (2005), Barkowsky et al. (2005). This thesis applies this method to spatial knowledge focusing on a basic problem of spatial cognition: **How to recognize variable spatial environments?**

1.4 The Assumption and the Criteria

Creatures interact with their surrounding environments by their sensors. Pigeons navigate by their magnetic sensors; dogs search foods by their noses; blind bats fly in the sky by their ultrasonic-wave sensors; rattle snakes catch foods by their infrared-ray sensors; people mostly use their eyes. Creatures with different sensors live in different conceptual spaces. The conceptual space of rattle snakes is based on the partition of temperatures of different objects in the surrounding physical space; their conceptual space shall not be as splendid as the conceptual space of people in the daytime, but they still can find foods without illumination, while people can not see anything without light.

The stimuli received by sensors change with the changes of the surrounding environment, the locomotion of the creature, even the malfunctions of the senors themselves. By looking at a clock, you will see different stimuli at any second, supposing that its second hand moves per second and its other hands move accordingly. However, you will have the same concept for all these different stimuli: It is the same clock. The pattern of your home with the door open is different from that with the door closed, however, they are patterns of the same environment.

The starting point of this thesis is that a cognitive agent perceives a snapshot view of spatial environments, recognizes objects from this snapshot view, and remembers a target spatial environment in the mind. It researches into the question of the representation of a snapshot view and the question of the comparison between the representation of the current perceived snapshot view and the representation in the mind.

People observe spatial environments and remember them. The first criterion for the representation of snapshot views is, therefore, that **the representation of spatial environments shall be acquirable through observation**. People are able to describe spatial environments. The second criterion is therefore that **the representation shall provide a systematical way to give meanings to spatial linguistic descriptions**. People may have distortions of spatial relations. Third criterion is therefore that **the representation shall provide ways to explain formations of spatial distortions**. People recognize familiar spatial environments, e.g., their homes, offices, quickly and easily. The criteria for the comparison of two representations with regard to recognizing spatial environment is therefore that **the process shall be computable (the decision process shall end in an amount of time that is polynomial in the size of the input)**.

1.5 Results and Contributions

Research results are as follows: Spatial relations between extended objects are understood by the connectedness relation with some extension objects — Distance relations between extended objects are understood by the connectedness relation between one extended object and the near extension region of the other extended object. Orientation relations between extended objects are understood by the connectedness relation between a near extension region of an extended object and a side of the other extended object. People have commonsense knowledge of the relative stability of spatial objects. A representation of spatial layouts is structured by objects and spatial relations following the principle of selecting reference objects that *the reference object is of higher relative stability than the location object and that the nearer objects are preferred to be reference objects*. The representation of a snapshot view is mereotopologically formalized. Recognizing a perceived snapshot view as target environment is interpreted as the compatibility between the representation of the perceived snapshot view and the representation of the target snapshot view.

The degree of the compatibility is determined by the spatial differences between the two representations and the relative stabilities of related objects. Recognizing spatial environments is formalized as a particular relationship between two representations of snapshot views.

This research work contributes to qualitative spatial representation by the commonsense understanding and the mereotopological formalism of spatial relations between extended objects. It contributes to artificial intelligence by the commonsense understanding and the formalism on the representation of snapshot view of spatial environments and the reasoning on recognizing variable spatial environments. It opens a new research direction in artificial intelligence/spatial cognition—vista spatial cognition.

1.6 The Organization of This Thesis

The thesis is structured as follows.

Chapter 2 presents the state of the art that pertains to recognizing variable spatial environments. I will review related work in cognitive psychology, object recognition, cognitive maps, formal spatial ontology, commonsense knowledge, and qualitative spatial representation.

Chapter 3 elaborates the research topics and research problems in detail based on the state of the art.

Chapter 4 presents the commonsense understanding of spatial relations, the relative stability of an extended object, relative spaces, the structure of snapshot views of environments, and the procedure of recognizing a spatial environment.

Chapter 5 formalizes the commonsense understanding in Chapter 4. It proposes a region-based representation and reasoning framework for recognizing spatial environments.

Chapter 6 introduces a symbolic computational system that pertains to the formalism in Chapter 5 — the LIVE model. The LIVE model is implemented in LispWorks 4.2 both on the Linux Susie 6.3 platform and on the Windows XP professional platform.

Chapter 7 summarizes and evaluates the results, and presents some future work.

Chapter 2
The State of the Art

In this chapter I review research work in cognitive psychology, object recognition, cognitive maps, formal spatial ontology, commonsense knowledge, and qualitative spatial representation that pertain to recognizing spatial environments.

2.1 Psychological Spaces

The structure of space can be described from the point of view of behavior.

(Piaget, 1954, p.212)

Montello (1993) classified four kinds of psychological spaces based on the way of observation: Figural space, vista space, environmental space, and geographical space. Vista space refers to a space that is projectively as large as the human body and that can be apprehended from one place without the need for locomotion. Typical vista spaces are single rooms, corridors, small valleys, etc. Tversky et al. (1999b) classified three kinds of psychological spaces: the space of the body, the space surrounding the body, and the space of navigation. The space of the body is *the space of our own actions and our sensations, experienced from the inside as well as the outside. It is schematized in terms of the natural parts of our body, our limbs;... the space surrounding the body is the space that can be seen from a single place, given rotation in place* (Tversky et al., 1999b, p.517). The space surrounding the body is conceptualized in three dimensions; the space of navigation is the space that is too large to be seen from a single place and it is conceptualized in two dimensions.

For out-door spatial environments, there might be differences between Montello (1993)'s vista space and Tversky et al. (1999b)'s space surrounding the body. When you see the moon at night, it is located in a vista spatial environment at that moment, however, it might not be very suitable to say that it is located in the space surrounding you, as it is quite far away from your body. The notion of vista spatial environments only emphasis the visual perception; while the notion of the space surrounding the body also emphasis actions of bodies or limbs besides the visual perception. For recognizing in-door spatial environments, what people see is located in a reachable distance, therefore, Montello (1993)'s "vista spatial environment" is

T. Dong: Recognizing Variable Environments, SCI 388, pp. 11–26.
springerlink.com

equivalent to Tversky et al. (1999b)'s "the space surrounding the body". "Spatial environments" in this thesis refer to Montello (1993)'s "vista spatial environment" and Tversky et al. (1999b)'s "the space surrounding the body".

2.2 Object Recognition (Figural Spatial Cognition)

Humans have a visual system that demonstrates remarkable object recognition ability, Liter and Buelthoff (1996). They can recognize objects very accurately and very fast. Research in cognitive psychology on human's object perception shows: (1) People tend to categorize objects at a preferred category; (2) people recognize objects based on some grouping laws (Gestalt theory); (3) there are two correlated models for object recognition: The view-dependent model and the view-independent model.

2.2.1 Object Recognition at the Preferred Category

> *The physical and social environment of a young child is perceived as a continuum. It does not contain any intrinsically separate "things". The child, in due course, is taught to impose upon this environment a kind of discriminating grid which serves to distinguish the world as being composed of a large number of separate things, each labelled with a name.*
>
> *(Leach, 1964, p.34)*

Humans' visual system is capable of recognizing objects from stimuli. Three common visual abilities of the human visual system are *discrimination*, *categorization*, *identification*.

The normal meaning of recognizing an object is that the object is successfully *categorized* into a particular object class, Liter and Buelthoff (1996). A *category* is a number of objects that are considered equivalent. A *taxonomy* is a system by which categories are related to one another by class inclusions. A level of abstraction within a taxonomy refers to a particular level of inclusiveness, Rosch et al. (1976). For example, if something belongs to the category of kitchen chair, it must also belong to the category of chair; if something belongs to chair, it must belong to the category of piece of furniture. Thus, in this example, we have a simple taxonomy which has three levels: The top is the category of furniture, next is the category of chair, the lowest level is the category of kitchen chair.

Identification means that an object belongs to a category that only has one object. Normal cases for identification are the human-face recognition and the environment recognition. When you recognize the face of your mother, your boss, or Bill Clinton, there is only one instance of your mother, your boss, or Bill Clinton. When you recognize the spatial environment of your current home, your current office, the main train station in Bremen, there is also one instance in the world to be your current home, your current office, or the main train station in Bremen. This thesis addresses the identification of a spatial environment, such as Mr. Bertel's home, your office, etc.

Discrimination means a yes-no judgment of whether two objects belong to the same category.

There is plenty of research showing that objects are recognized first at a particular level of abstraction, e.g., Rosch et al. (1976), Jolicoeur et al. (1984).

Rosch et al. (1976) argued that categories within taxonomies of concrete objects are structured such that there is generally one level of abstraction at which humans find it easiest to name objects and recognize them the fastest, namely "basic level category". Basic level of abstraction in a taxonomy is the level at which categories carry the most information, possess the highest cue validity, and are, thus, the most differentiated from others. Experiments in Rosch et al. (1976) showed that basic objects are the categories whose members: (a) Possess significant numbers of attributes in common; (b) have similar shapes; (c) are the first categorization made during perception of the environment, showing that basic objects have greater cognitive primacy than subordinate and superordinate categories.

Jolicoeur et al. (1984) proposed the notion of "entry point level", which means that every object has one particular level at which contact is made first with semantic memory. This level corresponds to the basic level in most cases. However, there are exceptions. For example, an expert airplane mechanic shifts the entry points of airplanes toward subordinate levels:

> *One subject ... a former airplane mechanic. His taxonomy was interesting. The lists of attributes common to airplanes produced by most subjects were paltry compared to the lengthy lists of additional attributes which he could produce. Furthermore, his motor programs as a mechanic were quite distinct for the attributes of the engines of different types of planes. Finally, his visual view of airplanes was not the canonical top and side images of the public; his canonical view was of the undersides and engines.*
>
> *(Rosch et al., 1976, p.430)*

2.2.2 Gestalt Grouping Laws

Gestalt psychology was started in Germany in 1912 by Max Wertheimer, Wolfgang Koehler and Kurt Koffka. The center tenet of Gestalt psychology is that the whole is more than the sum of its parts. The sound of a melody is much more than the sequence of its musical notes. If the melody is played one musical note after one musical note every one minute, people will be in difficulty in enjoying it. The same situation happens, when a film (a sequence of pictures) is played slower than 20 pictures per second. The Gestalt theorists maintained that it is the parts of the melody or the film that interact with one another and produce the nature of the whole, Rock and Palmer (1990).

To explain how human perceptions are formed, Gestalt theorists proposed the principle of "Praegnanz" (regularity) that perception tends to use the simplest and most regular organization to group the perceived stimuli, e.g., Koehler (1929), Koffka (1935), Wertheimer (1958). A partly occluded object will appear to continue behind an occluding object when such a continuation produces units that are more homogeneous in color or texture (grouping law of similarity), more smoothly contoured (grouping law of good continuation). Spelke (1990) summarized that young infants are sensitive to Gestalt grouping laws. For example, young infants can detect a misaligned contour in an array of elements with aligned contours. This shows that humans are sensitive to the Gestalt grouping law of good continuation at very young age.

2.2.3 View-Dependent vs. View-Independent Models

We can distinguish these five types of behavior : (1) . . . , (2) . . . , (3). . . , (4) the "reconstruction of an invisible whole from a visible fraction" (5) . . . .

(Piaget, 1954, p.14)

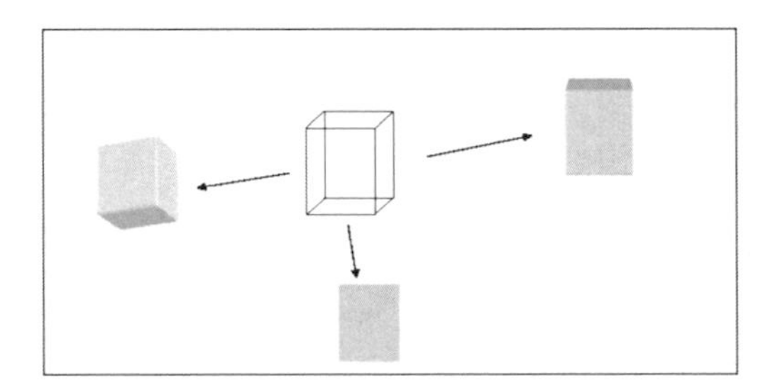

Fig. 2.1 Three kinds of sides of a cube

Intuitively, when we look around, we only perceive sides of spatial objects; when we perceive an object from different perspectives, different sides of the object are perceived. In Figure 2.1, we can see at most three sides of the cube from a viewpoint; when we perceive the cube from different view points, we can see different sides. There is a debate in the community of object recognition: The view-independent school proposes that the representation of a 3-dimensional object should be regardless of the viewpoint of the observer, e.g., Marr and Nishihara (1978), Biederman (1987), Marr (1982), Biederman and Gerhardstein (1993), and Biederman and Gerhardstein (1995). In contrast, the view-dependent school proposes that the representation of the 3-dimensional object is to represent multiple images of an object from different view-points, e.g., Breuel (1992), Buelthoff and Edelman (1992), Humphreys and Khan (1992), Tarr (1995), and Tarr and Buelthoff (1995).

Marr and Nishihara (1978) and Marr (1982) proposed a view-independent theory of object recognition. According to this theory, object recognition is carried out in two stages. In the first stage a view-dependent $2\frac{1}{2}$ sketch is formed; in the second stage a view-independent 3-dimensional object representation is formed from this $2\frac{1}{2}$ sketch.

Biederman (1987) proposed the theory of "Recognizing an object By its Components", namely RBC theory, which is built on Marr and Nishihara's early work. The assumption is that a modest set of generalized-cone components, named *geons*, can be derived from contrasts of readily detectable properties of edges in a two-dimensional image. In RBC theory, an object is represented by its components and the qualitative spatial relations among them, and recognizing an object is carried out by recognizing component(s) of the object.

On the other hand, there is plenty of research providing evidence that humans' object recognition performance is strongly viewpoint-dependent, e.g., Breuel (1992), Humphreys and Khan (1992), Buelthoff and Edelman (1992), Edelman and Buelthoff (1992), and Tarr (1995).

According to view-dependent approaches, object recognition depends on what is observed during familiarization with this object. Objects can be more readily recognized from some familiar orientations compared with others, Palmer et al. (1981). For example, people have more difficulty in recognizing the mixer when they have a bird-view compared to a field-view of the machine, shown in Figure 2.2.

Fig. 2.2 In (a) the mixer is viewed from top to down (a bird view); in (b) the same mixer is viewed normally (a field view). People are more easily to recognize the mixer with the field view (b) than the bird view (a). The picture is copied from (Biederman, 1987, p.144)

Humphreys and Khan (1992) conducted three psychological experiments to examine how novel 3-D objects are represented in long-term memory and how different views affect their representations. To address the issue of view dependency, subjects were trained to become familiar with objects in a specific orientation, then they were tested for recognizing these objects oriented in the familiar view and also novel views. The view-dependent approach and the view-independent approach make different predictions about the accuracy and time taken to recognize objects as a function of orientation: No significant difference between recognition of familiar views and unfamiliar views supports the view-independent approach; significant difference supports the view-dependent approach. The three experiments show that at least under certain conditions, the visual system stores a viewpoint specific representation of objects.

Buelthoff and Edelman (1992) showed that human recognition is better described by two-dimensional view interpolation than by methods that rely on object-centered three-dimensional models and that humans recognize objects best when they see them from a familiar view and worse from other views.

After debates between view dependent and view independent approaches in human object recognition, i.e., Biederman and Gerhardstein (1995) and Tarr and Buelthoff (1995), Tarr and Buelthoff (1998) reviewed findings from psychophysics, neurophysiology, machine vision, and behavioral results on the view-dependent / independent object recognition and concluded that although the view-dependent approach holds great promise, it has potential pitfalls that may be best overcome by structural information. And they proposed a hybrid approach for object recognition, which incorporates the most appealing aspects of view dependent and view independent approaches.

The RBC theory is, to some extent, something between a view-independent approach and a view-dependent approach. On the one hand, it suggests that to recognize an object is to recognize some of its components, which has some flavor of a view-dependent approach, as from different viewpoints, different components should be perceived. On the other hand, it inherits the view-independent approach of Marr and Nishihara to reconstruct the components from 2-D images.

Getting into the debate of the view dependent/independent issue is beyond the scope of this thesis. However, two general assertions are accepted by both of the two approaches: (1) An object can be recognized by parts of it; (2) there are structural relations among parts of an object.

2.3 Cognitive Maps, Frames, and Other Schemata

Physicians, engineers, mechanics and others use errors as signs of malfunctioning, that some system has broken down and is in need of repair. Not so for psychologists. Errors are viewed as natural products of the systems, and as such are clues to the way the system operates.

(Tversky, 1992, p.131)

In this section I review work on mental representations of spatial environments. These models have been given different names by different researchers. Trowbridge (1933) named them imaginary map; Shemyakin (1962) named them mental map; Appleyard (1969) named them environmental image; Boulding (1956) and Lynch (1960) named them spatial image; Lee (1968) named them spatial schema; Tolman (1948), Downs and Stea (1973), Kaplan (1973), and Kuipers (1975) named them cognitive map; Minsky (1975) named them frame; Tversky (1993) named them cognitive collage. In this section, I use *mental spatial representations* (MSR) to refer to all the above names in general. The following topics are reviewed: The existence of the represented world of MSR, the structure of MSR, the cognitive reference points in MSR, and the exploration of MSR from sentences.

2.3.1 The Existence of the Represented World of MSR

Tolman (1948) proved the existence of the represented world of mental spatial representations, namely cognitive maps, by serial experiments with rats. "Latent learning" experiments show that rats learn the maze structure which they run inside. "Vicarious trial and error" experiments show that rats can discriminate simple visual patterns and relate them with anticipations. "Searching for the stimulus" experiments show that rats often have to look actively for the significant stimuli in order to form its map and do not merely passively receive and react to all the stimuli which are physically present. "The hypothesis" experiments show that rats go through a maze by making systematic choices, which proved that rats have their "hypotheses" of spatial environments that results in their systematic choices (a rat may choose all right-hand doors, then all left-hand doors, then all dark doors). "Spatial orientation" experiments strikingly show that rats have higher level representations of spatial environments that allow them to re-orient themselves when the former path is blocked.

2.3.2 The Partial Hierarchical Structure of the MSR

The structure of MSR is addressed directly or indirectly in cognitive modelling, e.g., Siegel and White (1975), Stevens and Coupe (1978), Byrne (1979), Davis (1981), Tversky (1981), McDermott (1981), McNamara et al. (1984), Hirtle and Jonides (1985), McNamara (1986), Tversky (1991), McNamara (1991), McNamara (1992), and in artificial intelligence, e.g., Minsky (1975), Kuipers (1977), Yeap (1988), Yeap and Jefferies (1999), Barkowsky (2002), Barkowsky (2003), and Barkowsky et al. (2005).

Three classes of MSR theories exist in the literature: (1) non-hierarchical theory, e.g., Byrne (1979), which proposed that relations among spatial objects are mentally represented in a network; (2) strong hierarchical-theory, which proposed that spatial objects may group into "regions" leading to the hierarchical tree of regions and that spatial relations are not explicitly represented among regions at the same level in the hierarchical tree; (3) partial hierarchy-theory, e.g., Davis (1981), McDermott (1981), Kuipers (1978), and Stevens and Coupe (1978), which allowed spatial relations to be explicitly represented among regions at the same level in the hierarchical tree.

Byrne (1979) proposed that spatial relations among objects were mentally represented as a propositional network. For example, a mental representation of an urban environment could be a network of spatial objects with topological relations.

Stevens and Coupe (1978) and Tversky (1981) found that people make large systematic errors in judging spatial relations between two locations. For example, most people judge by mistake that San Diego (CA) is further east than Reno (NV) or that Madrid (Spain) is further south than Washington (DC) or that Seattle (USA) is further south than Montreal (Canada). To account for these errors, Stevens and Coupe (1978) proposed a partially hierarchical structure of mental spatial representation. That is, spatial relations between two locations are explicitly stored if the two locations are located in the same spatial region; spatial relations can be inferred by combining related spatial relations.

Another perspective of research that supports the existence of hierarchical structure is the influence of barriers to subjects' cognitive performance, e.g., Kosslyn et al. (1974), Cohen et al. (1978), Thorndyke (1981), and Newcombe and Liben (1982).

Kosslyn et al. (1974) found that children exaggerate distances between two locations that have either a transparent or an opaque barrier between them and adults make distance exaggeration when there is an opaque barrier. Newcombe and Liben (1982) found that subjects made exaggeration with rank order data. Cohen et al. (1978) found that cognitive maps of a familiar environment in 9-10 year old children and adults are strongly influenced by barriers, such as buildings, trees, and hills which tend to prohibit traveling. Distance estimation is based on the ease of travel. Distances are exaggerated when there are more barriers; distances are underestimated when there are less barriers. Barriers may influence the cognitive map by providing a method for chunking the space into subspaces which leads to the hierarchical structure of cognitive map.

Hirtle and Jonides (1985) used the Ordered Tree Algorithm, McKeithem et al. (1981), to show that people form a hierarchy of spatial objects on the basis of spatial and non-spatial attributes and that clusters in a hierarchy have consequences for performance in various tasks. McNamara (1986) systematically tested non-hierarchical and hierarchical schema with the results that partial-hierarchical schema theories are preferable.

In the AI community, Minsky (1975) proposed a partial theory of schema, namely the frame theory. A frame is a data structure consisting of nodes and relations which represents a situation. The top levels of a frame are fixed; the lower levels have many variables that can be assigned specific instances. Thus, a frame is also structured hierarchically. Kuipers (1975) represented the scenario of a cube world using the frame theory.

2.3.3 Cognitive Reference Points in the MSR

Wertheimer (1938) first suggested certain "ideal types" among perceptual stimuli used as anchoring points for perception. An introspective judgment by Wertheimer was: A line of 85° was almost vertical, but a vertical line was not almost 85°.

Rosch (1975) examined the existence of such "ideal types" in natural categories, which were named "cognitive reference points". She conducted psychological experiments for three domains: Colors, line orientations, and numbers. For the color system, "red", "yellow", "green", and "blue" are preferred cognitive reference points: The desaturated red was judged "muddy" by subjects but still named "red"; and the off-hue red was judged "purplish", but still "red" (Rosch, 1975, p.536). For decimal numbers, multiples of 10 are "ideal type" numbers. People say that "99.231 is around 100" rather than "100 is around 99.231". For line orientations, vertical, horizontal and diagonal lines are reference orientations.

In spatial cognition some spatial objects are used as cognitive reference objects to locate other objects. Sadalla et al. (1980) and Couclelis et al. (1987) investigated reference points in large scale spatial environments. Sadalla et al. (1980) found that landmarks are used to define the location of adjacent spatial objects and that subjective distances between reference points and non-reference points are therefore asymmetrical. Couclelis et al. (1987) found that landmarks may be discriminable features of a route, or discriminable features of a region, or salient information in a memory task. Locations in large spatial environments are partitioned into subregions each having a reference point.

2.3.4 Exploring the Structure of the MSR through Spatial Linguistic Descriptions

Talmy (1983) discussed how language is effective for conveying spatial information. He proposed that language schematizes space by selecting certain aspects of a reference scene to represent the whole, while discarding others. The schematization of indoor spatial environments in Ullmer-Ehrich (1982) discarded all small objects, like apples, cups, books, pens, etc. and only selected big ones. Foos (1980) investigated the construction of MSR of environmental spaces from language descriptions. The possibility of exploring the structure of MSR through linguistic descriptions is further strengthened by Schematization Similarity Conjecture that *to the extent that space is schematized similarly in language and cognition, language will be successful in conveying space* (Tversky and Lee, 1999, p.158).

2.4 Spatial Ontologies

2.4.1 SNAP and SPAN Ontologies

Grenon and Smith (2004) proposed two spatial ontologies, called SNAP and SPAN, to model different aspects of dynamic situations. SNAP and SPAN ontologies are partial in the sense that each is a window to the reality. SNAP is a snapshot ontology of endurants existing at a time. *SNAP entities have continuous existence in time, preserve their identity through change and exist in toto if they exist at all*[1]. SPAN is a four-dimensionalist ontology of processes. *SPAN entities have temporal parts; unfold themselves phase by phase and exist only in their phases*[2]. For example, the lobster, a nation, a population, an ocean are instances of SNAP ontologies; while the growth of a lobster, the history of a nation, the migration of a population, or the tide of an ocean are instances of SPAN ontologies.

---

[1] From http://www.spatial.maine.edu/actor2002/participants/smith-short.pdf

[2] See the above note.

2.4.2 *Fiat Boundaries and Fiat Objects*

One reason for resisting scepticism in face of the fiat world turns on the fact that people kill each other over fiat borders, and they give their lives to defend them.

(Smith and Varzi, 2000, p.405)

Boundaries can be either physical or non-physical parts of spatial entities. For example, the surfaces of planets or footballs are parts of these spatial entities. Such boundaries have material constitution. However, not all the boundaries have material constitution. For example, national borders or province borders are not based on the material discontinuities between two nations or two provinces; the equator of the earth and the North Pole are based on some mathematical measurement. Accordingly, two basic typologies of boundaries are distinguished: *bona fide* (or physical) boundary and *fiat* (in the sense of human decision or delineation) boundary, Smith and Varzi (2000). Two basic typologies of spatial objects are followed: *bona fide* objects which have *bona fide* boundaries and *fiat* objects which only have *fiat* boundaries.

Some *fiat* boundaries are dependent on *bona fide* boundaries. For example, every *bona fide* spatial object has the *fiat* boundary as its closure. When two people shake hands or kiss each other, they do not share a common part of their skin (*bona fide* boundaries) rather parts of their *fiat* boundaries "coincide". *Fiat* boundaries are boundaries which *exist only in virtue of the different sorts of demarcations effected cognitively by human beings* (Smith, 2001, p.135). They owe their existence both to acts of human decision (or *fiat*, to laws or political decrees, or to related human cognitive phenomena) and to real properties of the underlying factual material, Smith and Varzi (2000) and Smith (2001). The shores of the North Sea are *bona fide* boundaries; but we conceive the North Sea as a *fiat* object nonetheless, because "(its objectivity) is not affected by the fact that it is a matter of our arbitrary choice which part of all the water on the earth's surface we mark off and elect to call the 'North Sea' ", Frege (1884). The constellation "Great Bear" is a *fiat* object which owes its existence both to some huge stars (factual material) and to human cognitive activity (creation of a *fiat* boundary around these stars which has the shape of a bear), shown in Figure 2.3. Natural language makes a good contribution to the generation of *fiat* boundaries, namely "Linguistic Fiats" (Smith, 2001, p.141). For example, natural language expressions like "this" and "that" create an ephemeral *fiat* boundary in space, shown in Figure 2.4. *To set out the constraints on drawing fiat boundaries is a task that is by no means trivial* (Smith, 2001, p.138), because drawing *fiat* boundaries (or creating *fiat* objects) is based on a commonsense understanding of reality. Some *fiat* boundaries may not have sharp boundaries in reality. Consider the example of the Belgian enclave of Baarle-Hertog and its neighbor, the Dutch community of Baarle-Nassau, shown in Figure 2.5. The brighter areas represent the community of Baarle-Hertog. The small darker areas depict the tiny Dutch enclaves of Baarle-Nassau. Each such enclave is surrounded by a portion of Belgian territory, which is surrounded once more by Dutch territory (Smith, 2001, p.156).

Fig. 2.3 The *fiat* boundary around the stars of the Constellation "Great Bear"

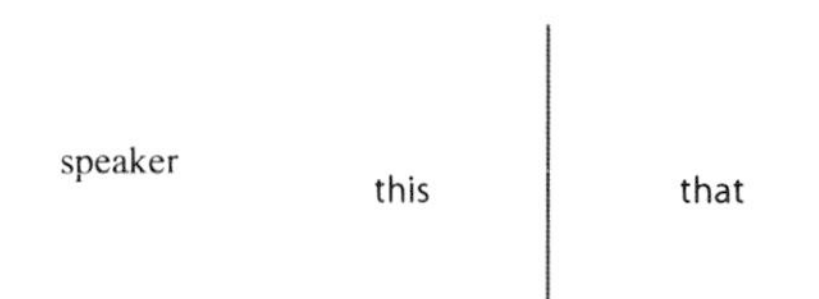

Fig. 2.4 Ephemeral *fiat* boundary established by the use of indexical terms (Smith, 2001, p.142)

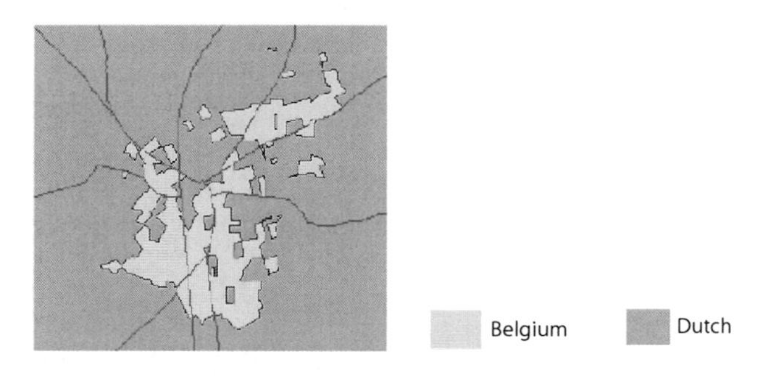

Fig. 2.5 The Enclaves of Baarle-Hertog and Baarle-Nassau. The picture is copied from (Smith, 2001, p.156)

2.5 Modelling Commonsense Knowledge

> *Naive Physics is the body of knowledge that people have about the surrounding physical world. The main enterprises of Naive Physics are explaining, describing, and predicting changes to the physical world.*
>
> *(Hardt, 1992, p.1147)*

After recognizing that most artificial intelligence systems in the late 1970s were toy systems, Hayes (1978) coined the term *Naive Physics* and proposed that researchers should concentrate on modelling commonsense knowledge.

Kuipers (1977) and Kuipers (1978) constructed a model of commonsense knowledge of large-scale space. Kuipers (1979) defined commonsense knowledge as follows: "...*(commonsense) knowledge about the structure of the external world that is acquired and applied without concentrated effort by any normal human that allows him or her to meet the everyday demands of the physical, spatial, temporal, and social environment with a reasonable degree of success*."

In the research field of Spatial Information Theory, Egenhofer and Mark (1995) coined the term *Naive Geography* that is concerned with formal models of the commonsense geographic world. It links the knowledge that people have about the surrounding geographic world and the formal representation and reasoning of that knowledge. Smith (1995) reviewed relations among commonsense knowledge, artificial intelligence, naive physics, physical science, etc. and set out the goal of developing a theory of the commonsense world.

2.6 Qualitative Spatial Representations

In this section I review some region-based qualitative spatial representation work that relates to the formalism presented in Chapter 5.

2.6.1 Classic Topological Relations

Smith (1994) proposed that topological relations are the most fundamental relations for cognitive science. Lynch (1960) used the metaphor of a rubber sheet for this fundamental relation in the sense that spatial entities can spread, shrink, twist, bend, etc. (like a rubber sheet) as long as the connection relations among them are preserved.

Formalizing the classic topological relation was done simultaneously and independently at Cohn's group, e.g., Randell et al. (1992), Cohn (1993), Gooday and Cohn (1994), Gotts (1994), Cohn (1995), Cohn and Gotts (1996), Bennett et al. (2000b), Bennett et al. (2000a), Bennett et al. (2002), Bennett (2003), Bennett (2004), etc. and at Egenhofer's group, e.g., Egenhofer (1989), Egenhofer (1991), Egenhofer and Sharma (1993), Egenhofer (1993), Egenhofer (1994), Egenhofer and Franzosa (1995), Egenhofer and Franzosa (1991), Egenhofer (2005), etc.

Cohn's school took the region rather than the point as the primitive unit in qualitative spatial representation and reasoning. They maintain that regions are more natural to represent indefiniteness that is *germane* to qualitative representation; spaces occupied by any real physical objects are always regions, not mathematical points; in common sense the word "point" means a small region rather than a real mathematical point, Cohn et al. (1997). The RCC theory was therefore developed for the topological relations between extended regions. The RCC-8 relations are shown in Figure 2.6.

Egenhofer's school took the relational algebra perspective and represented a region X by a pair of the boundary of X (in the notion of "δX") and interior of X (in the notion of "X°"). Topological relations between two regions are represented through the emptiness or non-emptiness of the intersection relations among their interiors, boundaries, and exteriors. The topological relation between two extended objects is represented by matrices shown in Figure 2.7.

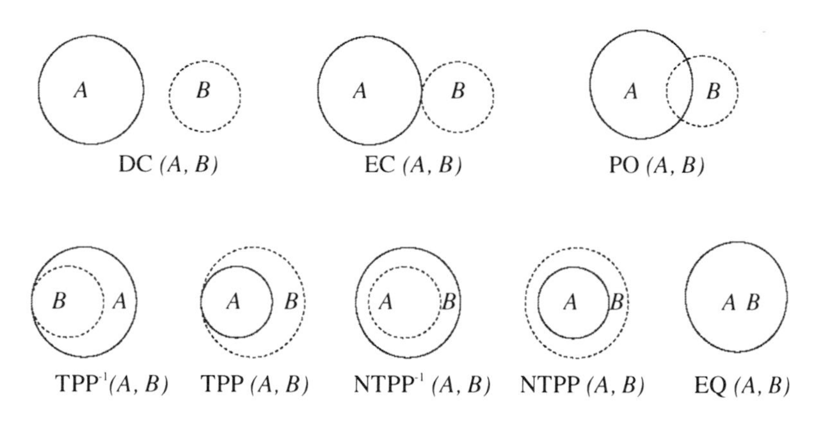

Fig. 2.6 The RCC-8 relations between regions

2.6.2 *Orientation Representations between Extended Objects*

Haar (1976) proposed a triangular model for the orientation relation between two extended objects. The space is partitioned into four mutually exclusive cones. The location of an object is represented by its centroid, which is the arithmetic mean of all the points of the object. The orientation relation is therefore determined by the centroid of the location object with regard to the four cones. However, the triangular model needs to be refined when the two extended objects are very close (or connected) or of irregular shapes, like the horseshoe-shape.

Guesgen (1989) extended Allen's (1983) 1-dimensional model into 2-dimensional and 3-dimensional models. In the extension to the 2-dimensional model, extended objects are projected into $x-$ and $y-$ axis. On each axis there is a 1-dimensional relation between the projections of the two objects. Spatial relations between two extended objects are represented by the pair of 1-dimensional relations. On each axis, nine orientation relations can be distinguished; therefore, the model can distinguish $9 \times 9 = 81$ relations for 2-dimensional space. As the model approximates relations of extended objects by projecting them to two 1-dimensional models, there are certain cases where this model has problems. For example, the two rectangles in Figure 2.8 are disconnected, however, their projections on the two axes are connected.

Goyal (2000) proposed the *coarse direction-relation matrix* which partitions space around an extended object and records into which tiles an extended location object falls. By counting in the ratio of the location object in each tile, the model can capture more detailed orientation information, shown in Figure 2.9.

Schmidtke (2001) assumed that direction information is unsuitable to be combined with topological information and formalized directional localization relations between extended regions. Her work introduced a system of sectors based on an extended object and the directional location of another object is described based on the sector system. The work yields a geometry of directional locations for 2-dimensional objects.

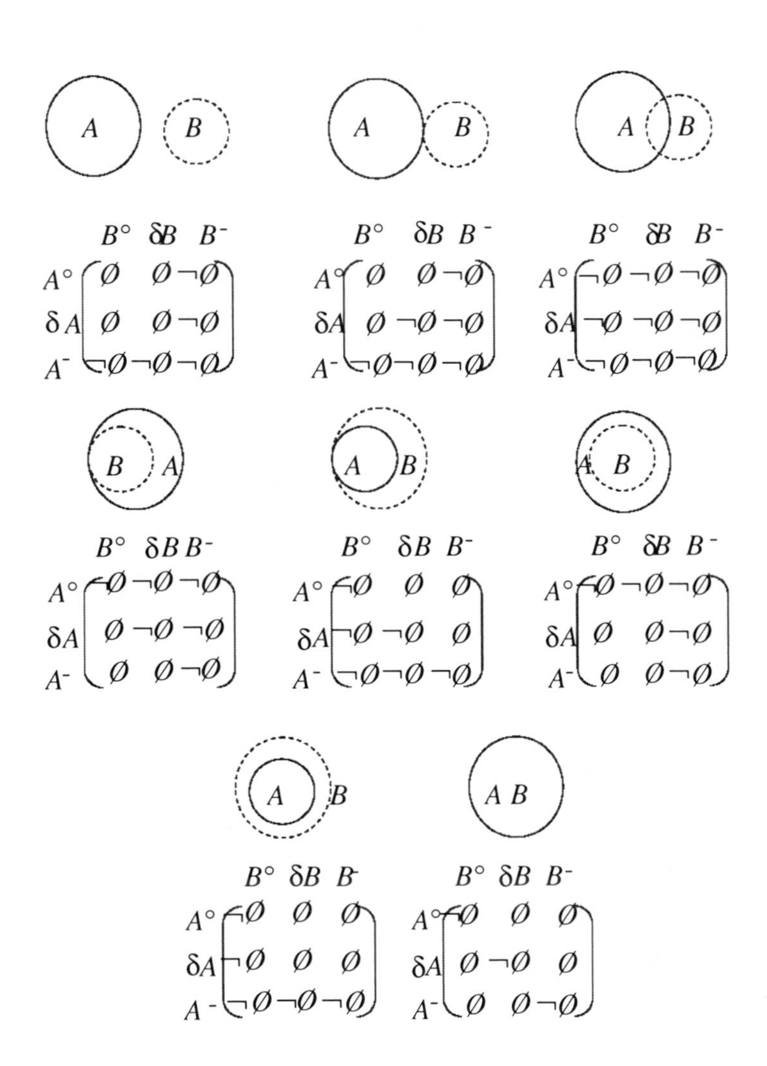

Fig. 2.7 The geometric interpretation of 8 topological relations between regions with the connectedness relation, Egenhofer (1994)

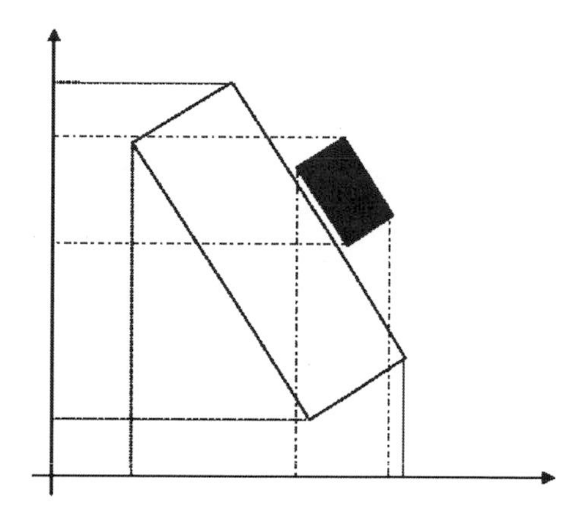

Fig. 2.8 The 'disconnected'-relation between the white rectangle and the black rectangle might not be suitably represented with the method in Guesgen (1989)

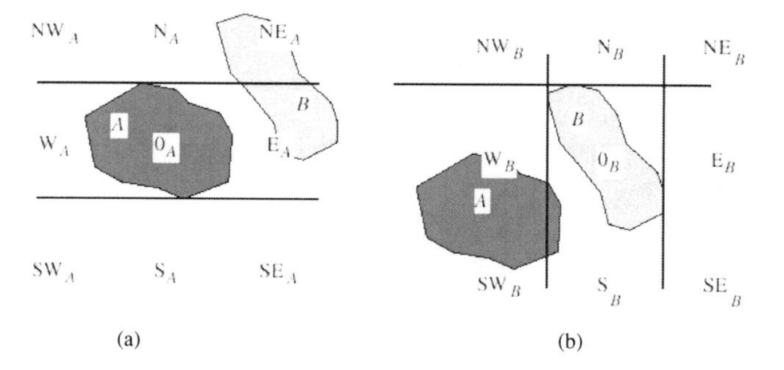

Fig. 2.9 The cardinal direction relation between two extended objects, A and B, is interpreted by the 'connected'-relation among the location object and the projection-based partitions of the reference object. The picture is copied from (Goyal, 2000, p.39)

2.6.3 Distance Representations between Extended Objects

Geometrical concepts between two extended objects–"solids" was defined by the primitive relation–"can connect" in de Laguna (1922). For example, region *A* is "longer than" region *B* is interpreted as: There are regions *W* and *X* such that *A* can connect *W* and *X* while *B* cannot; the distance between *A* and *B* is *zero*, if there is no *X* such that *X* cannot connect *A* and *B*. This work suggests that distance relations can be included into the topological framework.

There are also recent attempts trying to integrate all of the three systems of spatial relations into one framework. However, this has not proven successful, for example, Brennan et al. (2004) tried to propose a spatial ontology that brings together

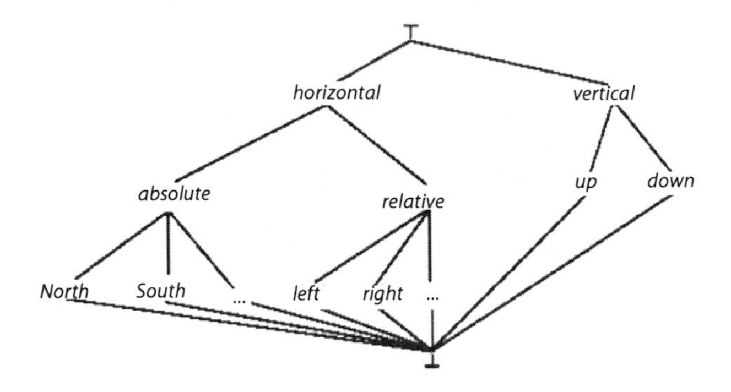

Fig. 2.10 The lattice definition of spatial orientations as shown in (Brennan et al., 2004, p.173)

three aspects of spatial knowledge, namely connectivity, proximity, and orientation. They attempted to model the "very close to each other" relation by splitting the "disconnected"(DC) relation of the RCC-8 theory into two relations: $DC_{=D}$ and $DC_{\neq D}$. However, the work ignores the "externally connected" (EC) relation in RCC-8. This results in the fact that $DC_{=D}$ is only a *façon de parler*[3] of EC. This approach abstracts extended objects as points, called "site", and introduces a pseudo-metric space with pseudo-distance D. Then for any site p, Brennan et al. (2004) proposed an influence area, denoted by $IA(p)$. To define the proximity relation, it introduced six kinds of "nearness" relations between a "site" and an "influence region" or between two "influence regions". Therefore, this part is still "point" based and not consistent with its connectivity part. In the orientation part, their work is even more problematic. Without providing ontological definitions of spatial orientation relations, the work presented a "lattice definition of the relation hierarchy of orientation relations", shown in Figure 2.10, and the composition of orientation relations of different reference systems equals to $\perp$, e.g., *left* $\wedge$ *North*$= \perp$. This obviously violates our common sense and research results in cognitive psychology: Tversky et al. (1999a) reported that people are likely to switch perspectives in describing orientation relations, therefore, there can be more than one orientation relation between two extended objects, such as *the courtyard is left of the church*, or *the courtyard is south of the church*, Lee (2002).

[3] The art of speaking.

Chapter 3
Research Topics and Research Questions

3.1 The Puzzle of Recognizing Environments

People receive features of stimuli from the surrounding environment, group them, and interpret them as objects. Recognizing objects means the perceived features can be grouped by a particular structure of features of an object (see 2.2.2). This follows that objects in the same category (grouped by the same structure of features) are equivalent and indistinguishable in isolation (see 2.2.1). Object categories are therefore multi-element categories. For example, people might have some difficulty in distinguishing twins. Figure 3.1 (left) shows the twin sisters from Taiwan: Sandy and Mandy. It would be quite difficult to identify which one is in the picture on the right.

(a) the twins: Sandy and Mandy

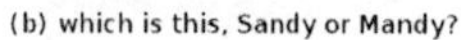

(b) which is this, Sandy or Mandy?

Fig. 3.1 People might have difficulty in distinguishing the twin sisters — Sandy and Mandy

Recognizing spatial environments, however, is different from recognizing single objects. Your home, your office, the entrance hall of your office building, etc. are of one-element category, although objects inside are of multi-element categories. That is to say that objects of multi-element categories come together and form an object of one-element category. How come? The environment as a whole is more than a set of all the objects inside. A spatial configuration is like a melody — the recognized objects and their correlations structure the configuration, just like musical notes and

T. Dong: Recognizing Variable Environments, SCI 388, pp. 27–32.
springerlink.com

their correlations structure a melody. Wilson et al. (1999) reported a memory impaired patient (LE), a sculptress suffered from autoimmune disorder[1]. This impairs her visual short-term memory[2] with mental image generation. She could not retrieve images from her memory, thus, she could only remember contours of objects. Consequently, she failed to distinguish two windows with different images in the church and she even had difficulty in recognizing the face of her husband[3]. However, she can locate objects; and amazingly, she can recognize her home. This case provides evidence that it is the relations among objects that turn a set of objects of multi-element categories into a configuration of one-element category. Two research topics are the knowledge of the snapshot view of a spatial environment (representation) and the use of the knowledge to recognize spatial environments (reasoning).

3.2 The Commonsense Knowledge of Spatial Environments

The knowledge of a snapshot view of an environment is called "Mental Spatial Representation" (MSR) (see 2.3.1–2.3.3). It can be explored through its external source — the perceived environment, and its product — the spatial linguistic descriptions. People perceive light features, e.g., colors, brightness, from the surrounding environment. They group the features together and interpret them as objects (see 2.2.2 and 2.2.3). For example, in Figure 3.2, the observer stands at P and faces the corner, so he can only perceive parts of the couch, parts of the door, parts of the walls, and parts of the tea-table. The observer can group stimuli from partial images by remembered features and recognize a couch, a tea-table, a door, and walls of the room. People select part of the objects as the components of the knowledge of the environment, while neglecting others, like the dust (see 2.3.4). However, the knowledge

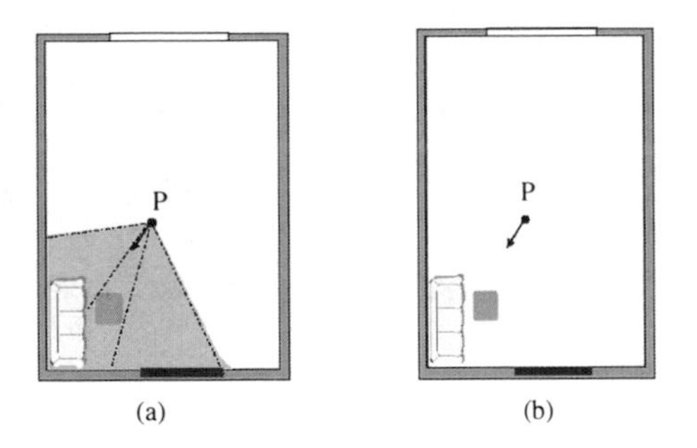

Fig. 3.2 In (a) the observer stands at P and faces to the corner. He perceives only part of the stimuli of the objects. However, he can recognize the couch, the tea-table, the door, and the walls of the room, shown in (b)

[1] Systemic lupus erythematosus.
[2] Dissociation between spatial span and pattern span.
[3] Personal communication with Allan Baddeley.

about objects in the environment is not the knowledge of the whole environment. Spatial linguistic descriptions (see 2.3.4) suggest that it shall include spatial structures among the objects. For example, in Figure 3.2, the observer may say, "the tea-table is in front of the couch; the couch is in the corner". The structure among objects includes spatial relations, such as "in front of" and "in", and the reference ordering, e.g., the tea-table is referenced to the couch, and the couch is referenced to the corner (see 2.3.3). The knowledge of a snapshot view of an environment is, therefore, at least composed of categorized objects and structural relations (including spatial relations and reference ordering relations) that are acquired through observation.

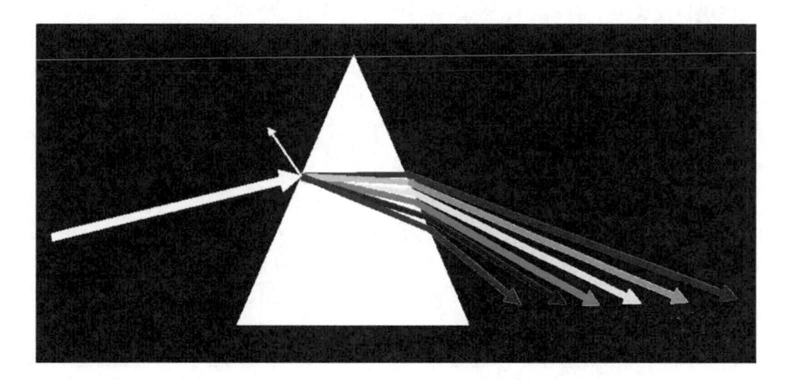

Fig. 3.3 When white light passes through a triangular optical prism, a spectrum will be formed

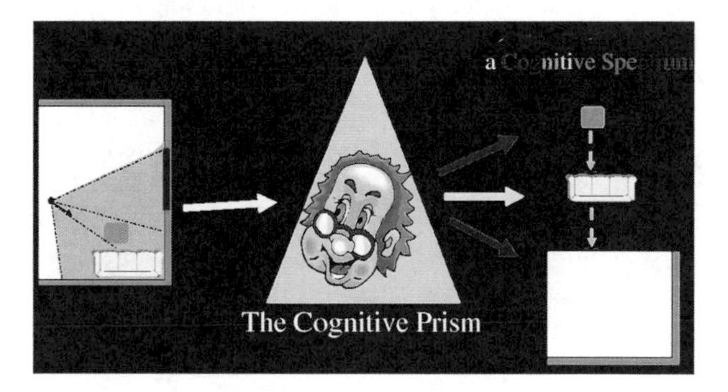

Fig. 3.4 When a scene passes through a cognitive system, a cognitive spectrum will be formed. Dotted arrows represent the ordering

When a beam of light reaches an optical prism, part of the light will be reflected, others will pass through, and light that passes through will be arranged with a particular ordering (refraction) and a spectrum will be formed, shown in Figure 3.3. When a scene enters the human eyes, some objects will be neglected, and the selected objects will be arranged with a particular ordering in the mind. The human cognitive system works like an optical prism which neglects some objects and re-arranges the selected by some properties, shown in Figure 3.4. It is, therefore, called "a cognitive prism" and the knowledge (the product of the cognitive prism) is called "a cognitive spectrum". It is a particular spatial configuration which is composed of objects, relations, and ordering.

The research questions are: (1) What are the spatial relations between extended objects through observation? (addressed in 4.2 and 4.3) (2) According to what property is the reference ordering formed? (addressed in 4.4 and 4.5) (3) What is the structure of a "cognitive spectrum"? (addressed in 4.6)

3.3 Recognizing Spatial Environments

When we talk about recognizing something, we talk about two worlds: One is the world before the eyes, namely, the perceived world; the other is the world in the memory, namely, the remembered world. Recognition means that the perceived world is believed to be the same as the remembered world.

If the remembered world is some remembered cognitive spectrums of a target environment, then these remembered cognitive spectrums can be transformed from one to another. Particularly, if they are temporally sequenced, then the first one and the last one span the target environment within a temporal duration, others in between demarcate traces of objects' movements during this temporal duration. From the perspective of spatial ontologies (see 2.4), the remembered cognitive spectrums are SNAP ontologies, and they participate into the SPAN of the target environment at different temporal points.

If the observer perceives an environment for some time, he will have a sequence of cognitive spectrums of the perceived world. Recognizing the perceived environment as the remembered one means that the perceived sequence of cognitive spectrums is believed to be those that follow the sequence of cognitive spectrums of the target environment that is remembered in the mind. That is, SNAPs of the perceived environment and SNAPs of the target environment participate into the same SPAN at two temporal durations — the remembered sequence is before the perceived one. This is equivalent to that the latest cognitive spectrum in the remembered sequence was before the first cognitive spectrum in the perceived sequence, and can be transformed to it. Because other cognitive spectrums in the remembered sequence transformed into the latest one and the first cognitive spectrum in the perceived sequence transforms into other ones in the perceived sequence, recognizing a spatial environment is equivalent to that a remembered cognitive spectrum is believed to transform into a perceived cognitive spectrum. One cognitive spectrum transforms into the other, if they are exactly the same, or the difference between them can be diminished by the replacements of the some objects in them. The ease of the replacements of related objects determines the degree of the confidence of the recognition result.

The research questions are: (4) How shall be two cognitive spectrums compared? (addressed in 4.7.1, 4.7.2, 4.7.3, and 4.7.4) (5) How does the difference between two cognitive spectrums effect the degree of the confidence of the belief that the perceived cognitive spectrum is transformed from the remembered one? (addressed in 4.7.5)

3.4 A Computational Approach to Recognizing Spatial Environments

To model the commonsense understanding of recognizing spatial environment, we will develop a computational theory and formalize spatial relations between extended objects. In most of the current representation models of distances and orientations, objects are either represented by points, e.g., Hernández (1994), Schlieder (1995), Zimmermann (1993) or by a directed segment, Freksa (1992). This is reasonable for geographical spaces and large-scale spaces: For geographical spaces the distances among stars or planets are much greater than the sizes of these objects, therefore, the point-based approach is suitable; for large-scale spatial environments locations are abstracted into points, thus, objects in the location are certainly abstracted into points, and objects in the street can be abstracted into a directed segment. However, for visa spatial environments (or the space surrounding the body), these geographical-oriented and large-scale-oriented point-based approaches have some problems.

Firstly, there is an incompatibility between "vista" and "point": To model a vista spatial environment is to model knowledge of something visible. The basic knowledge of visible extended objects is their shapes, colors, sizes, etc. If an extended object is represented by a point, then all the basic knowledge is missing, as the knowledge that a point represents is only an imagined extremely tiny mathematical location with a name, if such a location exists at all.

Secondly, there is an inconsistency of the treatment between spaces occupied by objects (occupied spaces) and spaces among them (unoccupied spaces). The two kinds of spaces are correlated and in the same granularity in vista spatial environments. People recognize occupied spaces, and the unoccupied spaces are the things that attach to the occupied spaces. Therefore, if there is a zooming system that zooms out occupied spaces into points, then it will also zoom out unoccupied spaces into points. This turns the whole vista environment into one point — this is a typical representation style for large-scale spaces.

Therefore, the first research question is: (1)' How can we formalize the spatial relations between extended objects through observation? (addressed in 5.2 and 5.3) In corresponding to the questions (2) to (5), other research questions in the computational part are: (2)' How can we formalize the reference ordering relation? (addressed in 5.4) (3)' What is the formal structure of a "cognitive spectrum"? (addressed in 5.5 and 5.6) (4)' How shall two formal structures of cognitive spectrums be compared? (addressed in 5.7.1-5.7.5) (5)' How to formalize the degree of the confidence of the belief that the perceived cognitive spectrum is transformed from the remembered one? (addressed in 5.7.6)

3.5 Towards the Theory of Cognitive Prism

Answering questions (1)–(5) and (1)'–(5)' results in a computational theory on recognizing variable spatial environments – The Theory of Cognitive Prism: When a cognitive system observes a spatial environment, it will select part of the objects in the environment, while neglecting others, and subjectively re-arrange the selected objects, forming a cognitive spectrum. To recognize a spatial environment, it compares the currently perceived cognitive spectrum with the cognitive spectrum of the target environment in its memory. The recognition result is determined by the ease of the transformation from the remembered cognitive spectrum to the perceived one.

Chapter 4
Recognizing Spatial Environments: A Commonsense Approach

This chapter presents a commonsense approach to recognizing variable spatial environments. It is structured as follows: Section 4.1 presents the starting point; section 4.2 proposes spatial relations between extended objects; section 4.3 presents the notion of the relative space; section 4.4 proposes the notion of the relative stability; section 4.5 proposes the principle of reference between extended objects; section 4.6 presents the structure of a cognitive spectrum; section 4.7 presents the relations between two cognitive spectrums that pertains to recognize spatial environments; section 4.8 summaries The Theory of Cognitive Prism.

4.1 Knowledge about Extended Objects Based on Observation

When Mr. Bertel's mother stood at the entrance door of her son's apartment, she saw view-dependent images of the objects — some partially blocked by others. However, she can recognize objects based on parts of one or more sides of them.

4.1.1 Preferred Categories of Objects Based on Observation

Recognizing objects means categorization. To recognize objects is to put them into the preferred categories (see 2.2.1). This owes both to real properties of the underlying factual material and to acts of human decision. When you see a chair near your kitchen table, you have a recognized chair in your mind; when you see a dog in a picture, you have an imagined dog in your mind; when you see a film, you have imagined spaces based on the sequence of the pictures. Your recognizing the chair near your kitchen table owes to the light reflection from the chair and to your recognition activity. Your recognizing the dog in the picture owes to the color distribution in the picture and to your knowledge of a dog in the mind. Imagined or recognized objects are, therefore, *fiat* objects (see 2.4.2) and exist in the mind. They sleep somewhere in the memory, and are awaken either by some external stimuli or by certain mental desires.

T. Dong: Recognizing Variable Environments, SCI 388, pp. 33–53.
springerlink.com

4.1.2 Sides of Recognized Objects

People see sides of objects from different perspectives and recognize them (see 2.2.3). Sides are distinguished and named qualitatively, such as the *left* side, the *right* side, the *back* side, the *upper* side, the *bottom* side. Sides are parts of the boundary of an extended object. Your face is the *front* side of your head. Your two ears are located on the *left* side and the *right* side of your head. The boundaries of the *front* side, the *left* side, and the *right* side are *fiat*, as there are no physical discontinuities between the *front* side and the *left* side, and between the *front* side and the *left* side of your head. Neighborhood sides are partially overlapped as when you are seen from the front, parts of your *left* side and your *right* side will also be seen. When you are seen from the left, part of the your *front* side will be seen.

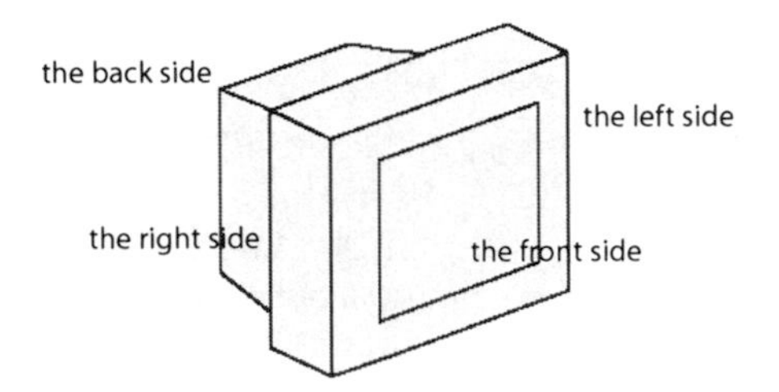

Fig. 4.1 Different sides of a TV set

4.1.3 Spatial Relations among Sides

Sides of an object have some simple spatial relations. When you walk clockwise from the *front* side to the *back* side of the TV set, the surface that you have passed is the *right* side of the TV set. When you walk clockwise from the *back* side to the *front* side of the TV set, the surface that you have passed is the *left* side of the TV set, shown in Figure 4.1. Sides of an object have neighborhood relations. The *front* side is a neighbor of the *left* side and the *right* side; the *left* side is a neighbor of the *front* side and the *back* side; etc.

4.2 Spatial Relations as Spatial Extensions

When Mr. Bertel's mother stood at the entrance door, she could recognize not only objects in the apartment, but also spatial relations among them. Imagine that she stood at the place and faced the apartment as shown in Figure 4.2 (a), she would recognize spatial relations between herself and objects, such as *she is nearer to the balloon than to the writing-desk, the balloon is in front of her*, and also spatial relations among objects, such as *the balloon is near and in front of the writing-desk.*

If you turn on a flashlight, it will emit a light beam. Imagine our eyes are such a flashlight that they can emit a light beam and that we see the side of the object

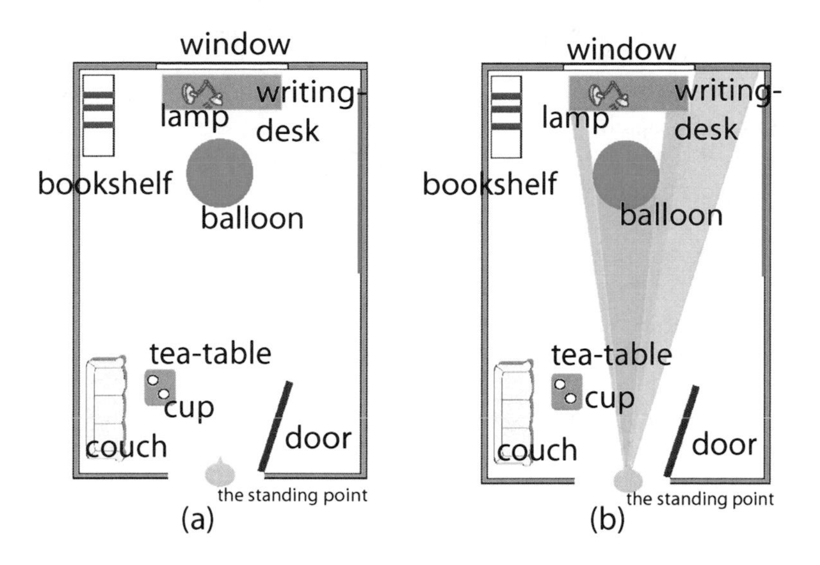

Fig. 4.2 Spatial relations based on observation

which blocks the light beam[1]. Then, *that an object is in front of the observer* can be explained as follows: "If the observer faces to the object, there will be a light beam which is connected with both the eyes and the object", shown in Figure 4.2 (b); *that an object is left to the observer* can be explained as: "If the observer turns to the left, there will be a light beam which is connected with both the eyes and the object", or "if the observer faces forward, there will be a light beam which is connected with the left side of the observer's face and the object" — The observer can prove this by turning to the left to see whether there is a light beam connecting the object and the eyes; *that the observer is nearer to object A than to object* can be explained as: "The light beam which is connected with the observer and *object A* has a smaller size than the light beam which is connected with the observer and *object B*".

Imagine the observer does not emit light beams, rather ultrasonic waves, like blind bats, or imagine a blind man with a stick, they can also know whether there are obstacles in front of them and which obstacle is nearer to them than other objects, if they know the size of the emitted ultrasonic wave or the stretching degree of the stick. Light beams, ultrasonic waves, and sticks, etc. serve as extensions of the body space of the observer, no matter what these extension objects are, no matter these extension objects are created actively or received passively. Spatial relations between the observer and objects can be understood by making extension of the body space into the space surrounding the body.

[1] This is the way that the ancient Greece explained how people see objects. Modern physics explains that it is not that eyes can emit light, rather eyes can receive reflection light from the objects.

By assigning extensions to objects, people can define new spatial relations between objects. As people experience that they can reach the writing-desk while sitting on the chair, they could assign the space of their body to the chair and indicate the spatial relation between the chair and the desk as *the chair is near the writing-desk*. If they notice that they reach the front side of the writing-desk while sitting on the chair, they would give the orientation relation between the chair and the desk as *the chair is in front of the writing-desk*.

Three natural extension objects relating to recognize spatial environments are: The body (or part of the body, like limbs), the step(s)[2], and the light beam (in physics it is called the *reflection light*). When Mr. Bertel's mother stood at the entrance door of her son's apartment, she recognized the objects using light beams, and knew the spatial relations among objects. If one object can be extended to the other by the body, then the two objects are near; if the extension needs several steps, the distance between them would be a bit far away; if the extension is only possible by the light beam, the two objects would be far away.

4.2.1 Distances: The Extension from One Object to the Other

When Mr. Bertel's mother observed that she was nearer to the chair than to the writing-desk, she just used the light beam as extension objects, and she found that the chair blocked part of the light beam which is connected with the writing-desk and the eye, shown in Figure 4.2 (b). That is, the light beam that is connected with her eyes and the chair has a smaller size than the light beam that is connected with the eyes and the writing-desk. This is also proved by the fact that if she walked to the writing-desk following the light beam which is connected with the writing-desk and the eye, she would reach the chair first. This time she used steps as the extension objects. Distances between two extended objects are the degree of the extension from one object to the other by certain extension objects.

4.2.2 Orientations: The Extension to Which Side

If the observer sits in the chair in Figure 4.3, she/he could easily reach the front side of the desk. That is, the chair with the body of the observer as the extension is connected with the front side of the desk. The spatial relation between the chair and the desk is not only that *the chair is near the desk* but also *the chair is near to the front side of the desk, instead of its other sides*. That is, the chair is nearer to the front side of the desk than to its other sides. This is described as an orientation relation that *the chair is in front of the desk*. The orientation relation between two extended objects can be explained by the distance comparison between one object and sides of the other object.

Some objects are perceived the same from different perspectives, e.g., the black ball and the white ball in Figure 4.4 (a). An observer can describe the orientation

[2] Steps represent a sequence of body spaces such that successive body spaces are connected at legs.

Fig. 4.3 Qualitative orientation relations among the cup, the chair, and the desk will be described as *the cup is on the right side of the desk* and *the chair is in front of the desk*

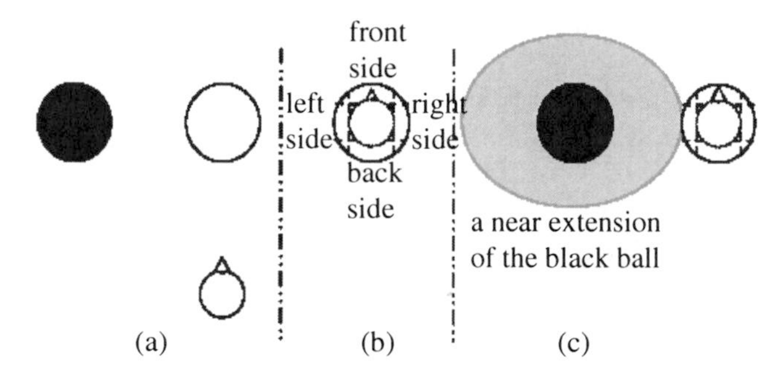

Fig. 4.4 A *fiat* projection of the observer to the white ball

relation between the two balls as follows: *If I stand at the place shown in Figure 4.4 (a), the black ball is at the left hand side of the white ball.* This statement assigns a name to the side of the white ball which is nearer to the black ball than its other sides. This side is named based on the standing place and the facing direction of the observer. The observer imagines that she/he were at the white ball while keeping the facing direction and names sides of the white ball with reference to names of her/his own sides, shown in Figure 4.4 (b). Then, the orientation relation between the black ball and the imagined herself/himself can be given: "The black ball is located left to the imagined her/him". In the language description, the observer says: "*If I stand at the place shown in Figure 4.4 (a), the black ball is at the left hand side of the white ball.*" It is such a mechanism as if the observer "projects" herself/himself to the white ball, and this mechanism is called "the *fiat* projection". This mechanism requires the observer to know the standing place and the facing direction of herself/himself and to extend this knowledge to the space surrounding the body[3]. From the perspective of the *fiat* projection mechanism, the deictic orientation reference

[3] This is consistent with the empirical findings of Franklin and Tversky (1990) and the functional spaces, i.e., the space of the body, the space around the body, in Tversky et al. (1999b) and Tversky (2005).

framework in the literature is a kind of naming the sides of an object[4]. In Figure 4.3, the orientation relation between the cup and the desk can be explained by the *fiat* projection as follows: The observer imagines herself/himself sitting in the chair, and projects her/his sides to the desk, then the cup is nearer to her/his right side. The linguistic description can be: "*The cup is on the right side of the desk*".

4.3 The Relative Spaces

Space is simply the order or relation of things among themselves.

–Leibnitz

When Mr. Bertel stretches out his hand to the cup, his arm serves as the extension object of the body. He will say "the cup is near me", if he can hold the cup. If he holds the cup and moves the arm around, all the locations of the cup are near him. Therefore, "the cup is near me" means that the cup is located in one of the locations that all are "near" Mr. Bertel. If he stretches out the arm and holds a book, a cigarette, he will agree that the book or the cigarette is near him. Even he stretches out the arm and holds nothing, he will say, "there is nothing near me". Therefore, the existence of the space delineated by "near me" is independent of the existences of the cup, the book, the cigarette, etc. It is a **relative space** (following Smith (2001) it is a *fiat* object) created by Mr. Bertel and the arm. It refers to the space that Mr. Bertel can reach by the arm. Mr. Bertel is called the anchor object of the relative space; the arm is called the extension object of this relative space. Mr. Bertel's relative space of "near me" might have a different size from his mother's relative space of "near me" in the sense that their arms might be of different sizes. Mr. Bertel's relative space of "near me" is also changed along with the size of his arm. When he was a child, he had a smaller space of "near me" than the one that he now has.

In particular, a **connectedness relative space** is constructed by an anchor object such that any object in this relative space is connected with the anchor object. For example, the lamp is in the connectedness relative space of the writing-desk.

A **distance relative space** is a relative space constructed by an anchor object and one or more extension object(s) such that it spatially extends the anchor object to reach other objects by the extension object(s). For example, suppose that 'near' is interpreted as the distance relation such that two objects are 'near', if the body of the observer can connect the two objects, then 'near the couch' is a distance relative space such that any object in it can be connected with the observer's body which is also connected with the couch.

[4] Similarly, the absolute orientation reference framework can be explained as follows: The observer projects the imagined earth to the reference object, so that its sides could be named after the imagined earth: The north, the west, the south, and the east. The absolute orientation relation is therefore the result of the distance comparison between the location object and the four imagined sides of the reference object. In general, the *fiat* projection is described as follows: The observer projects an imagined object which has clear sides to the reference object. The orientation relation is the result of the distance comparison between the location object and the imaged sides of the reference object.

An **orientation relative space** is a relative space constructed by an anchor object with its particular side such that any object inside this relative space is nearer to this particular side than to its other sides. For example, ‘in front of the desk’ is an orientation relative space constructed by the desk and its front side such that if Mr. Bertel is in this relative space, he will be nearer to the front side of the desk than to its other sides.

Objects are recognized categorically. This leads to the fact that relative spaces are also recognized categorically. The category of a relative space is determined by the categories of the objects and relations that construct the relative space. In particular, the category of a connectedness relative space is determined by the connectedness relation and the category of its anchor object. The category of a distance relative space is determined by the distance relation and the category of its anchor object. The category of an orientation relative space is determined by the orientation relation and the category of its anchor object.

4.4 Knowledge of the Relative Stability

People have the common sense that the ground and the sky are like a motionless box and that other objects are inside, moving from here to there. This box with the ground as the bottom and the sky as sides and the cover is more stable than the objects inside, because when a box moves, the objects inside are moved along with it, however, when the objects inside move, the box does not move along with them. This implies the principle of commonsense reasoning of the relative stability between two objects as follows: *Object A is relatively more stable than object B, if object B moves along with the move of object A, and object A does not move along with the move of object B*. For example, a table is relatively more stable than a picture on the table, because the picture moves along with the move of the table, but not vice versa.

In the common sense, the earth is motionless and people frequently move here and there on the earth. Trees and lawns are planted in the ground; streets, plaza, and buildings are built on the ground. They are almost as motionless as the earth. For indoor spatial environments, floors, walls, and ceilings are motionless components of buildings. Objects that are affixed to these motionless components are as stable as them, e.g., sinks in the kitchen. All of the motionless objects along with objects that are affixed on them are called *rarely moved objects*. Big pieces of furniture, such as writing-desks, couches, shelves, etc., are moved into rooms. Therefore, they are relatively less stable than these *rarely moved objects*. On the other hand, people walk actively inside of rooms. Objects that are often held by people move along with them, such as newspapers, cups of tea, glasses of wine, books, pens, shoes, plates, etc. These objects are called *always moved objects*. Big pieces of furniture can not be held easily by people, therefore, people move themselves from one big piece of furniture to another along with *always moved objects*. For example, Mr. Bertel takes a book from the bookshelf, then walks to the desk; later, he walks to the couch and reads newspapers while drinking a cup of tea. Big pieces of furniture may have peripheral objects to support people or some *always moved objects*. Couches have tea-tables as the peripheral objects which support cups, glasses, apples, newspapers, etc.; desks and tables have chairs as the peripheral objects that support people. Peripheral objects, therefore, have relatively stable locations with corresponding big pieces of furniture. And, it is usual that they are moved by people for other usages. For example, Mr. Certel moved the balloon from near the desk to near the tea-table to chat with Mr. Bertel. Therefore, they are less stable than the big pieces of furniture. As a summary, the qualitative space of the relative stability is bounded by two poles — the motionless pole (the earth) and the motion pole (people) and spanned according to the principle of commonsense reasoning of the relative stability as follows: *Rarely moved objects* are closest to the motionless pole, such as walls, floors, sinks in the kitchen, etc.; followed by the *seldom moved objects*, such as writing-desks, couches, etc. *Always moved objects* are closest to the motion pole (people), such as cups, shoes, newspapers; *often moved objects* are located between the *seldom moved objects* and *always moved objects*, such as chairs, tea-tables, etc. *Rarely moved objects* are relatively more stable than *seldom moved objects*; *seldom moved objects* are relatively more stable than *often moved objects*, and so on. The diagrammatical representation is shown in Figure 4.5.

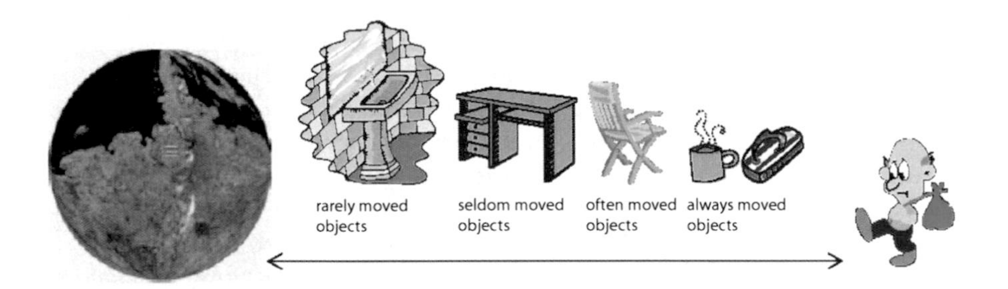

Fig. 4.5 Four object classes of indoor spatial environments based on the relative stability

When you move your arm, when you open your mouth, when you close your eyes, your body may not move. The arm, the mouth, and the eyes are functional parts of the body. When a functional part of an object moves, the object itself may not move. However, when your front side moves, your body must move along with the front side. It is therefore assumed that the object and its sides have the same relative stability.

4.5 The Reference Ordering

4.5.1 The Principle of Reference in Spatial Linguistic Descriptions

When a pilot has lost his location information, he expects the location information such as "*you are above the South Pole*" rather than "*you are in your plane*", because the latter provides no information about the location of the plane. His relative location should be referenced to more stable objects. The commonsense knowledge of the relative stabilities affects the selecting of the reference object in spatial linguistic descriptions, in order to keep the description informative. For example, in describing relative locations of *the sun*, people prefer to saying that "*the sun is in the sky*" and "*the sun moves around the earth*" instead of "*the sky surrounds the sun*" or "*the earth moves around the sun*"[5]. People prefer to saying "*the bike is beside the tree*", "*the bike is on the lawn*", "*the car is on the plaza*", "*the carpet is on the floor*", "*the picture is on the wall*", instead of "*the tree is beside the bike*", "*the lawn is under the bike*", "*the plaza is under the car*", "*the floor is under the carpet*", "*the wall is behind the picture*". This is summarized as the first criterion of selecting reference objects in spatial linguistic description as follows: The reference object in spatial linguistic descriptions should be of the higher or the same relative stability than the object. This is called "the criterion of stability" which keeps the linguistic description informative.

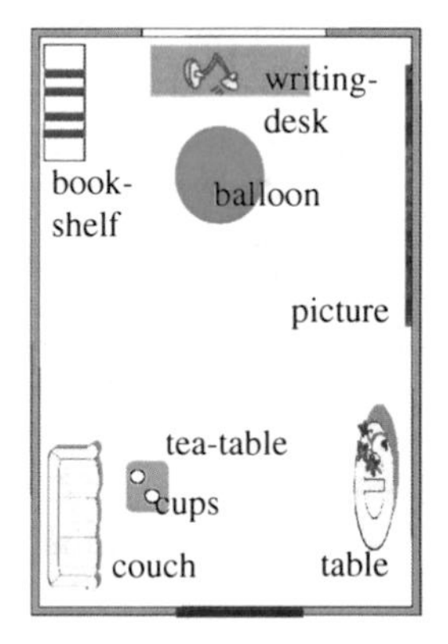

Fig. 4.6 The cups are referenced to the tea-table; the balloon is referenced to the writing-desk; the picture is referenced to the walls; etc.

According to the criterion of stability, the writing-desk in Figure 4.6 can be referenced to the door — it is 'far away from the door'. However, 'the door' is normally not selected as the reference object, because it is farther away from 'the writing-desk' than 'the window' and 'the front wall' which also satisfy the criterion of stability. This infers that among reference objects satisfying the criterion of stability, the nearer reference objects are preferred, because the farther away a reference object is, the more effort will be cost to make spatial extensions from the reference object to the object. This is called "the criterion of economics". In Figure 4.6, there are many objects that are relatively more stable than the cups, however, 'the tea-table' is selected as the reference object for the cups, instead of 'the couch', 'the

[5] In the history, people who insisted saying that *the earth moves around the sun* were even burnt to death.

floor', 'the ceiling', because according to the criterion of economics, objects that are connected with the cups have the priority to be the reference objects. The tea-table is connected with the cups; and it is also of higher relative stability than cups, which satisfies the criterion of stability.

The principle of reference in spatial linguistic descriptions is summarized as follows: The reference object in spatial linguistic description is of higher or the same relative stability than the location object (the criterion of stability) and the nearer reference objects are preferred (the criterion of economics).

4.5.2 The Cognitive Reference Objects

People have cognitive reference points (see 2.3.3) in spatial cognition. Following Rosch (1975)'s definition of "cognitive reference point[6]", to be a "reference object" in a vista spatial environment, an object must be shown to be one which other objects are seen "in relation to". For the task of recalling and describing spatial environments, "in relation to" is taken to mean, in the spatial linguistic descriptions, that the relationship between reference objects and non-reference objects are asymmetrical; whereas relationships between two non-reference objects are symmetrical. This follows that an object is less stable than its cognitive reference object. For example, in Figure 4.6, 'the writing-desk' is the cognitive reference object of 'the lamp', because people would say that *the lamp is on the writing-desk*, rather than *the writing-desk in under the lamp*; 'the bookshelf' is not the cognitive reference object of 'the writing-desk', because people would say that *the writing-desk is near the bookshelf* and that *the bookshelf is near the writing-desk*. The principle of selecting cognitive reference objects can be derived from the principle of reference in spatial linguistic description as follows: The cognitive reference object is of higher relative stability than the location object and that the nearer objects are preferred to be cognitive reference objects.

A relative space is defined as *a location* of an object, if the anchor object of the relative space is one of the cognitive reference objects of the object. For example, *that the tea-table is near the couch* is interpreted as *a location* of the tea-table is *near the couch*, because the couch is one of the cognitive reference objects of the tea-table. *The location* of an object in a spatial environment is defined as the conjunction of all its *locations*. For example, *locations* of the tea-table in Figure 4.6 are *on the floor*, *near the couch*, *in front of the couch*, then *the location* of the tea-table is the conjunction of the locations as follows: *on the floor* and *in front of the couch* and *near the couch*. Locations of objects delineate 'where' objects are in spatial environments.

[6] *To be a "reference point" within a category, a stimuli must be shown to be one which other stimuli are seen "in relation to." ... For purpose of the present research, "in relation to" was taken to mean, operationally, that there were judgment tasks in which the relationship between stimuli in the category and the reference stimulus were asymmetrical; ..., relationships between two non-reference stimulus members of the category were symmetrical.* (Rosch, 1975, p.532)

4.6 Cognitive Spectrums of Spatial Environments

When a snapshot view of a spatial environment passes through people's eyes, it is partitioned and understood as objects, and spatial relations, which are structured based on *the principle of selecting cognitive reference objects*, forming a cognitive spectrum of the spatial environment. For example, when a scene, shown in Figure 4.7, is perceived by an observer, it will be partitioned and understood as objects, such as a couch, a tea-table, a right wall, a front wall[7], and a floor. The observer has commonsense knowledge of relative stabilities of objects as follows: The floor, the front wall[8], and the walls are rarely moved objects, the couch is a seldom moved object, the tea-table is an often moved object. Based on the principle of selecting cognitive reference objects, objects shall be referenced to more stable objects nearby. Therefore, the cognitive reference objects of tea-table are the floor which is connected with the tea-table, and the couch which is near the tea-table; similarly, the cognitive reference objects of the couch are the floor, the right-wall, and the front-wall. As the floor and the walls are sides of the room, spatial relations among them are here beyond the scope. They are used as single objects, as evidenced in spatial linguistic descriptions.

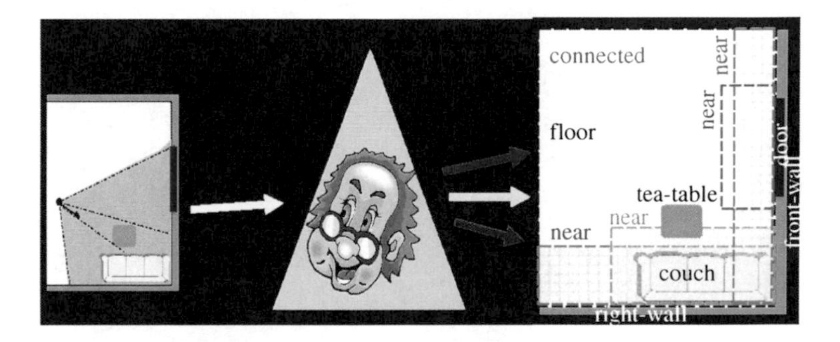

Fig. 4.7 When a scene passes through a cognitive prism, a cognitive spectrum will be formed

4.6.1 A Diagrammatic Representation of Cognitive Spectrums

A cognitive spectrum can be diagrammatically represented by nested relative spaces. Colors (or textures) of objects represent relative stabilities. An object is connected with a relative space, if it is referenced to the anchor object of this relative space and it has the spatial relation to the anchor object which is delineated by this relative space. The cognitive spectrum in Figure 4.7 can be diagrammatically represented by the nested relative spaces shown in Figure 4.8. The gray (cotton), green

[7] The names of the walls are based on the *fiat* projection of the observer who is facing the door. The right-wall is the side of the room on the right-hand side, the front-wall is the side of the room to which the observer faces.

[8] Although a door often turns on one side of the doorframe, the front wall with the door and the doorframe as a whole is a side of the room which is a rarely moved object.

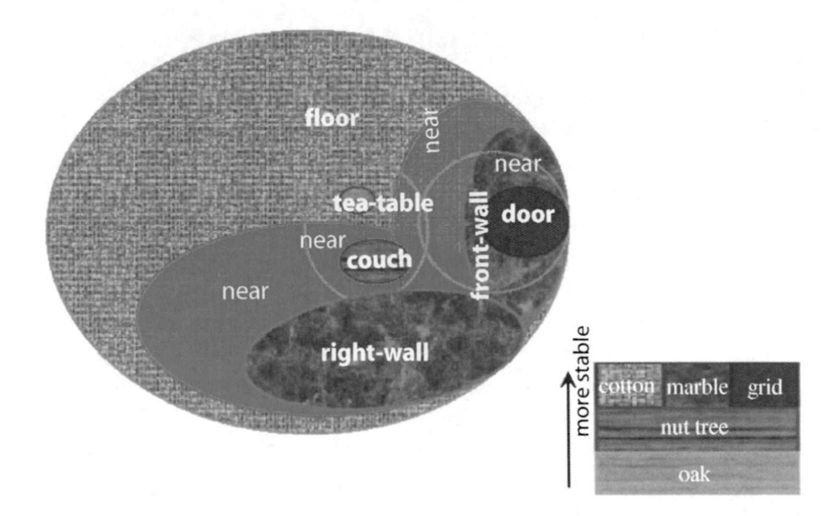

Fig. 4.8 A diagrammatic representation of a cognitive spectrum. The gray (cotton), green (marble), and blue (grid) represent rarely moved objects; the sandy beige (nut tree) represents seldom moved objects; the amber color (oak) represents often moved objects

(marble), and blue (grid) colors (textures) represent rarely moved objects; the sandy beige (nut tree) represents seldom moved objects; the amber (oak) represents often moved objects. Sandy beige (nut tree) objects can reference to gray (cotton), green (marble), and blue (grid) objects; amber (oak) objects can reference to Sandy beige (nut tree), gray (cotton), green (marble), and blue (grid) objects. A colored (textured) region with an object name represents the connectedness relative space of this object; a light colored region with a label 'near' around an object region represents a 'near' relative space such that objects connected with this relative space and disconnected with its anchor object are near the object region. Among the objects that the tea-table (the amber/oak region) can reference, the floor (the gray/cotton region) and the 'near' relative space of the couch (the sandy beige/nut tree region) are connected with the tea-table. Among the objects that the couch can reference, the floor, the 'near' relative space of the right-wall (the green region/marble), and the 'near' relative space of the front-wall (the green region/marble) are connected with the couch.

4.6.2 A Symbolic Representation of Cognitive Spectrums

Symbolically, a cognitive spectrum of an environment can be represented by two tables — the table of objects, which lists properties of objects in the environment, and the table of relative spaces, which lists locations of objects in the environment.

The table of objects collects object knowledge about the cognitive spectrum. The object knowledge includes the object name, its category, and its relative stability. For example, object knowledge of the scene in Figure 4.7 includes knowledge of

Table 4.1 The table of knowledge of objects

Name	*Category*	*Relative stability*
a room	$<\text{ROOM}_{Bertel}>$	rarely moved
a couch	$<\text{COUCH}_{Bertel}>$	seldom moved
a tea-table	$<\text{TEATABLE}_{Bertel}>$	often moved

a room, a couch, and a tea-table. The room is a rarely moved object; knowledge of the room characterizes a categorization $<\text{ROOM}_{Bertel}>$ which means objects in this category are indistinguishable from the room through perception; the couch is a seldom moved object; knowledge of the couch characterizes a categorization $<\text{COUCH}_{Bertel}>$ which means objects in this category are indistinguishable from the couch through perception. The tea-table is an often moved object; knowledge of the tea-table characterizes a categorization $<\text{TEATABLE}_{Bertel}>$ which means objects in this category are indistinguishable from the tea-table through perception.

Knowledge of object sides provides feature to recognize an object. Object sides can be named through the *fiat* projection of the observer, like the front side of the writing-desk, or the salient side of the object itself, like the floor, the ceiling, the wall with the door. The knowledge of object sides, as part of knowledge of object category, is not explicitly listed in the table of knowledge of objects, rather it is implied in the knowledge of the object category.

Table 4.2 The structure of the table of relative spaces

	a spatial relation, e.g., in front of
an object, e.g., *the couch*	objects located in this relative space, e.g., *the tea-table*

The table of relative spaces represents locations of objects in the environment. The left column is the list of cognitive reference objects; the top row of the right column is a cell for a spatial relation. Then a cognitive reference object and a spatial relation delineates a relative space which is the cell locating at the same column as the spatial relation and the same row as the cognitive reference object. Object names in this cell represent objects that are referenced to the cognitive reference object at this row and located in this relative space. The structure of the table of relative spaces is shown in Table 4.2.

An object in the cell of a connectedness relative space of its reference object represents the object such that it is relatively less stable than the anchor object of the connectedness relative space and that it is connected with the anchor object. For example, in Figure 4.8 objects in the connectedness relative space of the room are the couch and the tea-table, shown in Table 4.3, because they are less stable than the room and they are connected with the room.

An object in the cell of a distance relative space represents the object such that it is relatively less stable than the anchor object of the distance relative space and that

Table 4.3 The table of connectedness relative spaces

	connected with
the room	the couch, the tea-table

it has a distance with the anchor object. For example, in Figure 4.8 objects in the near relative spaces are listed in Table 4.4.

Table 4.4 The table of distance relative spaces

	near
the couch	the tea-table
the right wall	the couch
the front wall	the couch

An object in the cell of an orientation relative space represents the object such that it is relatively less stable than the anchor object of the orientation relative space and that it is nearer to one side of the anchor object than to its other sides (denoted by the orientation relation). For example, the tea-table in front of the couch is located in the orientation (front) relative space of the couch, shown in Table 4.5, which means the tea-table is nearer to the front side of the couch than to its other sides.

Table 4.5 The table of orientation relative spaces

	in front of
the couch	the tea-table

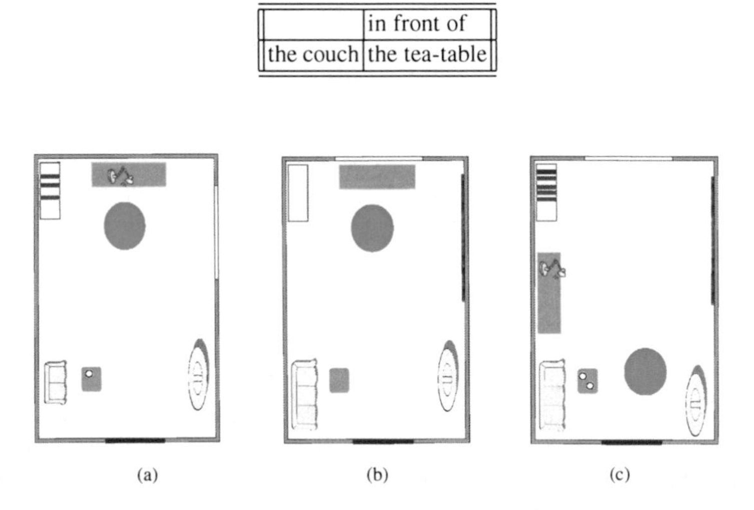

Fig. 4.9 Mr. Bertel's mother looked at Mr. Certel's apartment (a) and *mapped* it with her target layout — Mr. Bertel's apartment (b). She concluded that it was not Mr. Bertel's apartment. At last she found her son's apartment (c)

4.7 Relations between Two Cognitive Spectrums

Unfortunately, Mr. Bertel's mother was neither able to recognize Mr. Certel's apartment by recognizing Mr. Certel's objects, nor able to recognize his son's apartment by recognizing his object. The reason is that recognizing objects is only categorization. Two objects in different snapshots are indistinguishable through perception, if they are of the same category. When Mr. Bertel's mother visited her son's apartment for the third time (on the ninth day), she went into Mr. Certel's apartment by mistake, Figure 4.9 (a), while Mr. Certel was not in the apartment. Although the layout was very similar to Mr. Bertel's, she knew that it was not her son's home and left the apartment. She knew this because she had a cognitive spectrum of Mr. Bertel's apartment in the mind. She knew what objects should be in the apartment and where the objects should be, etc. When she went to Mr. Certel's apartment, the cognitive spectrum of Mr. Bertel's apartment in the mind was used as the criterion to justify whether the perceived one was the target apartment. She expected a window in the front wall[9], however, there is no window in the perceived front wall; she expected a picture on the right wall[10], however, there is no picture but a window on the perceived right wall. She expected a big couch in the corner of the perceived apartment, however, the couch in the perceived apartment is smaller. She knew that sides of rooms, locations of windows, and sizes of couches can not be changed easily. This led to the doubt whether it was Mr. Bertel's apartment. When she came to the apartment shown in Figure 4.9 (c), she saw furniture that looked the same as those in Mr. Bertel's apartment; and she found that the window, the bookshelf, the couch and the picture were located in the expected places and that the writing-desk and the balloon were not located in the expected places. However, this environment was much more like Mr. Bertel's apartment, because she thought that all these differences could happen. Recognizing spatial environment is a judgement on the spatial differences between the cognitive spectrum of the currently perceived environment and the cognitive spectrum of the target environment. Following questions are raised: How shall two cognitive spectrums be compared? How shall the judgement be made?

4.7.1 The Categorical Comparison

The starting point to compare two cognitive spectrums is that objects of the same category are indistinguishable in isolation through perception. The primitive relation between two cognitive spectrums is the *categorically the same* relation between two objects. Two objects are *categorically the same*, if they are of the same category. For example, Mr. Bertel's writing-desk and Mr. Certel's writing-desk are

[9] The names of the wall is based on the *fiat* projection of the observer who is standing at the door. The front-wall is the side of the room that the observer faces.

[10] The name of the wall is based on the same *fiat* projection as the above note. The right-wall is the side of the room on the right-hand side.

categorically the same, though they are different ones. However, Mr. Bertel's room and Mr. Certel's room are not *categorically the same*, because their windows are located in difference walls.

Two clusters of objects are *categorically the same*, if for each object category, there is the same number of objects in the two clusters. For example, the cluster of Mr. Bertel's writing-desk and his balloon and the cluster of Mr. Certel's writing-desk and his balloon are *categorically the same*, because for each object category there is one object in each cluster.

Two relative spaces are *categorically the same*, if they are of the same category. In particular, two connectedness relative spaces are *categorically the same*, if their anchor objects are *categorically the same*. For example, the connectedness relative space of Mr. Bertel's writing-desk and the connectedness relative space of Mr. Certel's writing-desk are *categorically the same*, because the two writing-desks are *categorically the same*. Two distances relative spaces are *categorically the same*, if their anchor objects are *categorically the same* and their distance relations are *categorically the same*. That two distance relations are *categorically the same* means that their extension objects are *categorically the same*. For example, the *near* relative space with Mr. Bertel's writing-desk as the anchor object and Mr. Bertel's mother as the extension object and the *near* relative space with Mr. Certel's writing-desk as the anchor object and Mr. Bertel's mother as the extension object are *categorically the same*. Two orientation relative spaces are *categorically the same*, if their anchor objects are *categorically the same* and their orientation relations are *categorically the same*. That two orientation relations are *categorically the same* means that the two orientation relations have the same name[11]. For example, the orientation relative space with Mr. Bertel's writing-desk as the anchor object and the front orientation relation and the orientation relative space with Mr. Certel's writing-desk as the anchor object and the front orientation relation are *categorically the same*.

Two clusters of relative spaces are *categorically the same*, if for each category of the relative spaces, there is the same number of relative spaces in the two clusters. Two conjunctions of relative spaces are *categorically the same*, if the two clusters of the components of the conjunctions are *categorically the same*. For example, let the first conjunction of relative spaces be *near the writing-desk in Mr. Bertel's room* and *on the floor in Mr. Bertel's room*, the second conjunction of relative spaces be *near the writing-desk in Mr. Certel's room* and *on the floor in Mr. Certel's room*, then the first conjunction and the second conjunction are *categorically the same*, because the two clusters of their components are *categorically the same*. That is, the cluster of *near the writing-desk in Mr. Bertel's room* and *on the floor in Mr. Bertel's room* is *categorically the same* as the cluster of *near the writing-desk in Mr. Certel's room* and *on the floor in Mr. Certel's room*.

[11] It assumes that they are based on the same *fiat* projection (this makes the two anchor objects have the same side names).

4.7.2 *The Process of Mapping Cognitive Spectrums*

When Mr. Bertel's mother stood at the door of an apartment, judging whether it was her target apartment, she should map objects in the target cognitive spectrum with objects in the cognitive spectrum of the perceived environment. Suppose that in her target cognitive spectrum there was a writing-desk and that she perceived two writing-desks that are *categorically the same* with the target writing-desk. Which of the two should be *mapped* to the target one? This depends on the location of the target writing-desk. If it was located in front of the window, then the one in front of the window in the perceived environment, if existed, was *mapped* to it. However, this explanation is not total, as it assumes that the target window was *mapped* to the perceived window. That is, she should map the target window to a window in the perceived environment before mapping writing-desks. This also depends on locations of the windows. If the target window was in the front wall, then the perceived window in the front wall was *mapped* to it. And again, this requires to compare the target front wall with the perceived front wall. At last, it requires to compare the two rooms at the very beginning.

This leads to the conditions of object mapping and the principle of the sequence of mapping objects between the target cognitive spectrum and the perceived one as follows: Cognitive reference objects shall be *mapped* before the mapping of their location objects. That is, rarely moved objects are firstly compared and *mapped*, followed by seldom moved objects, and often moved objects. For indoor spatial environment, firstly the perceived room is compared with the target room. If they are not *categorically the same*, then the perceived environment shall not be the target environment, else the two rooms are *mapped*[12] and each relative space with the perceived room or its side as the anchor object is *mapped* to a relative space with the target room or its side as the anchor object with the condition that the two relative spaces are *categorically the same*. For example, the room in Figure 4.9(a), namely $room_1$, and the room in Figure 4.9(b), namely $room_2$, are not *categorically the same*, because the window in the perceived room is located differently as that in the target room. So, the perceived cognitive spectrum is not the target cognitive spectrum. The room in Figure 4.9(c), namely $room_3$, and $room_2$ are *categorically the same*, therefore, $room_3$ and $room_2$ are *mapped*, and a relative space with $room_3$ as the anchor object will be *mapped* to a relative space with $room_2$ as the anchor object, if the two relative spaces are *categorically the same*. The connectedness relative space with $room_3$ as the anchor object is *mapped* to the connectedness relative space with $room_2$ as the anchor object; the near relative space with the front wall of $room_3$ as the anchor object is *mapped* to the near relative space with the front wall of $room_2$ as the anchor object; etc. For short, two rooms are *mapped*, if they are *categorically the same*. Two relative spaces are *mapped*, if their anchor objects are *mapped* and

[12] The location of a room is *mapped* when the observer is outside of the room, e.g., in the corridor. People's going into the room implies that the location of the room is *mapped* with the location of the target room and that the outer side of the perceived room is *categorically the same* as the outer side of the target room. It is therefore assumed that the locations of the rooms are *mapped*.

they are *categorically the same*. Two objects inside the rooms are *mapped*, if they are *categorically the same* and their locations are *mapped*. The process of object mapping proceeds by mapping objects in the *mapped* relative spaces as follows.

(Step 1) If the perceived room and the target room are *categorically the same*, then they will be *mapped* and put them as a pair into a queue; set the index to the first element of the queue.

(Step 2) If the index points to a pair in the queue, then take the pointed pair of the queue and find their *mapped* relative spaces, else stop;

(step 3) For each *mapped* relative spaces, find *mapped* objects and put each *mapped* objects as a pair to the tail of the queue;

(step 4) increase the index by 1, and go to (Step 2).

The *mapping* process produces a queue of object pairs. Two objects in each pair are *categorically the same* and *mapped* in location.

4.7.3 The Spatial Difference

The spatial difference is the un-mapped objects after the process of object mapping. If Mr. Bertel's mother saw a washing-machine in the perceived environment (in the connectedness relative space of the perceived room), and on the other hand, there was no washing-machine in the cognitive spectrum of the target environment, then the existence of this washing-machine cannot be *mapped* to an object in the target environment; if Mr. Bertel moved his writing-desk to Mr. Certel's room, and Mr. Bertel's mother saw an environment with no writing-desk, on the other hand, there was a writing-desk in the cognitive spectrum of her target environment, then the writing-desk in the target cognitive spectrum cannot be *mapped* to an object in the cognitive spectrum of the perceived environment. In general, for each object category in the two *mapped* relative spaces, if there are more objects of this category in the perceived relative space than those of the same category in the target relative space, new objects of this category are perceived, this kind of difference is called an *appearance*; if there are more objects of this object category in the target relative space than those in the perceived relative space, some objects of this category are not perceived, this kind of difference is called a *disappearance*. For example, there is a small couch in the perceived cognitive spectrum in Figure 4.9(a), and there is no such a small couch in the target cognitive spectrum in Figure 4.9(b). It is called *there is an appearance of a small couch in the perceived environment*.

4.7.4 The Compatibility

If we recognize that the perceived cognitive spectrum as the target cognitive spectrum, spatial differences between the two cognitive spectrums will be explained as the result of the transformations of the un-mapped objects. The ease of transformations can be determined by the relative stabilities of the un-mapped objects. The compatibility between the cognitive spectrum of the perceived environment and the target cognitive spectrum qualitatively represents the ease of the transformations.

The cognitive spectrum of the perceived environment and the target cognitive spectrum are hardly compatible, if there is any spatial difference that happens to *rarely moved objects*, such as rooms (including windows, doors), sinks. For example, the cognitive spectrum of the environment shown in Figure 4.9(a) and the cognitive spectrum of the environment shown in Figure 4.9(b) is hardly compatible, because the two rooms are not *categorically the same*, as there is a window in the right wall of the perceived environment, and there is no windows in the right wall of the target environment.

The cognitive spectrum of the perceived environment and the target cognitive spectrum are possibly compatible, if there is no spatial difference that happens to *rarely moved objects* and there is spatial difference that happens to *seldom moved objects*, such as writing-desks, couches, bookshelves, etc. For example, there is no spatial difference that happens to rarely moved objects between the cognitive spectrums of the environments shown in Figure 4.9(b) and Figure 4.9(c), and there is an appearance of a writing-desk in the near relative space of the left wall and a disappearance of a writing-desk in the near relative space of the front wall in the cognitive spectrum of the environment shown in Figure 4.9(c), writing-desks are seldom moved objects, therefore, the cognitive spectrum of the environment shown in Figure 4.9(b) and the cognitive spectrum of the environment shown in Figure 4.9(c) are possibly compatible.

The cognitive spectrum of the perceived environment and the target cognitive spectrum are compatible, if there is no spatial difference that happens to *rarely moved objects* and *seldom moved objects* and there is spatial difference that happens to *often moved objects*, such as chairs, tea-tables, sitting-balls, etc.

The cognitive spectrum of the perceived environment and the target cognitive spectrum are very compatible, if there is no spatial difference that happens to *rarely moved objects*, *seldom moved objects*, and *often moved objects* and there is spatial difference that happens to *always moved objects*, such as books, cups, pens, etc.

The cognitive spectrum of the perceived environment and the target cognitive spectrum are indeed compatible, if there is no spatial difference found.

4.7.5 Recognition as the Judgment on the Compatibility

Recognizing spatial environments is a judgment on the compatibility between the cognitive spectrum of the perceived environment and the cognitive spectrum of the target environment. In everyday life situation, people's activities in indoor spatial environments are limited to chatting, eating, playing, sleeping, walking, etc. These activities result in many spatial differences that happen to *always moved objects*, such as books, cups, pens, etc., some differences of *often moved objects*, such as chairs, sitting-balls, etc., a little bit differences of *seldom moved objects*, such as writing-desks, couches, etc., and hardly differences of *rarely moved objects* such as door frames, window frames, floors, ceilings, sinks, etc. So, people can recognize the perceived environment as the expected one according to the degree of the compatibility between the cognitive spectrum of the perceived environment and the cognitive spectrum of the target environment.

4.7.5.1 *No, It Is Hardly ...*

If the perceived environment is hardly compatible with the expected one, then in everyday life situation, the perceived environment will be hardly thought of as the expected one. For example, when Mr. Bertel's mother went to Mr. Certel's apartment, shown in Figure 4.9(a), and found that the window was located in a different wall, then she thought that it was hardly Mr. Bertel's apartment, and went out.

4.7.5.2 *It Might Be ...*

If the perceived environment is possibly compatible with the expected one, then in everyday life situation, the perceived environment might be thought of as the expected one. In this situation, the observer would ask for reasons for the spatial differences. For example, after Mr. Bertel's mother went out from Mr. Certel's apartment, she went to Mr. Bertel's apartment which showed her the after party layout, shown in Figure 4.9(c). She found that the writing-desk was located between the bookshelf and the couch and that the table and the balloon were located differently as she expected, then she wondered for a while, before recognizing it as Mr. Bertel's apartment. When she met Mr. Bertel, she asked what happened last night.

4.7.5.3 *It Is ...*

If the perceived environment is compatible with the expected one, then in everyday life situation, the perceived environment is thought of as the expected one. In this situation, the observer will take it for granted that it is the expected environment, and some observers may move these often moved objects back to the expected location. For example, when Mr. Bertel's mother went to Mr. Bertel's apartment for the second time (on the seventh day), she found that the perceived layout was compatible with the expected one and that the balloon was located differently. She mumbled why Mr. Bertel moved the balloon to the tea-table, and she moved the balloon back to the writing-desk.

4.7.5.4 *Yes, It Is Exactly ...*

If the perceived environment is very compatible with the expected one, then in everyday life situation, the perceived environment is thought of as exactly the expected one. In this situation, the observer will also take it for granted that it is the expected environment.

4.7.5.5 *Yes, It Is Indeed ...*

If the perceived environment is indeed compatible with the expected one, then in everyday life situation, the perceived environment is thought of as indeed the expected one. In this situation, the observer will certainly take it for granted that it is the expected environment. In most cases, such an indeed compatible layout will be changed by the activities of the people, for example, by reading newspapers people change the location of newspapers; by drinking tea people change the location of cups.

4.8 The Theory of Cognitive Prism

By answering questions (1)–(5) in Chapter 3, The Theory of Cognitive Prism is summarized as follows: When a cognitive system observes a spatial environment, it will select part of the objects in the environment, while neglecting others, subjectively re-arrange the selected objects based on the commonsense knowledge of relative stabilities, make spatial extensions from one object to others, and form a cognitive spectrum following the principle of selecting cognitive reference objects. To recognize a spatial environment, the cognitive system compares the currently perceived cognitive spectrum with the cognitive spectrum of the target environment in its memory following the stable-object-first order. Rooms are *mapped*, if they are *categorically the same*; objects inside rooms are *mapped*, if they are *categorically the same* and their locations are *mapped*. The compatibility between two cognitive spectrums is determined by the un-mapped objects in both cognitive spectrums and their relative stabilities. In everyday life situation, the recognition result is directly interpreted by their compatibility.

Chapter 5
The Formalism: A Region-Based Representation and Reasoning of Spatial Environments

This chapter formalizes the commonsense knowledge represented in Chapter 4. Formulae have the form of Z notion, Woodcock and Davies (1996), "$(\forall\, x|p \bullet q)$" (it is read as "for all x satisfying p, q holds"), "$(\forall\, x \bullet q)$" (it is read as "for all x, q holds"), "$(\exists\, x|p \bullet q)$" (it is read as "there is x satisfying p such that q"), "$(\exists\, x \bullet q)$" (it is read as "there is x such that q"), and '$\iota x(q)$' (it is read as "the x that q's"). x is the bound variable, p is the constraint of x, and q is the predicate. '`true`' stands for T and '`false`' stands for F. The proofs of the theorems in this chapter are listed in the appendix.

5.1 The Object Region and Its Properties

An *object region* represents an imagined or recognized object. An object region has a *fiat boundary* and its interior part. An object region belongs to the preferred category in which it is imagined or recognized; it has a degree of the relative stability. The preferred category of an object region corresponds to a collection of features which each object region in this category has. An object region has sides. A *side* is a part of the *fiat* boundary of the object region that can be seen from a view point. It is called a *side region*. It is assumed in (4.1.2) that neighborhood sides are overlapped, therefore, neighborhood *side regions* are overlapped. It is assumed in (4.3) that an object moves along with its sides, therefore, an object region and any of its side regions have the same relative stability. The side of an object region belongs to a category, which corresponds to a subset of the features of the preferred category of the object region.

Object regions are denoted by *MATHIT* capital letters, such as *DESK*, *CHAIR*, *DOOR*, ... Categories are denoted by `typewriter` capital letters, such as `DESK`, `CHAIR`, `DOOR`, ... Let *OBJ* represent an object region, and the preferred category of *OBJ* be `OBJ`, *OBJ*`.category` represents the preferred category of *OBJ*, *OBJ*`.category=OBJ`. *OBJ*`.stability` represents the relative stability of *OBJ*. For objects of indoor spatial environments, four degrees of relative stabilities are distinguished: *rarely moved*, *seldom moved*, *often moved*, and *always moved*. The

T. Dong: Recognizing Variable Environments, SCI 388, pp. 55–78.
springerlink.com

four degrees are represented by `rarelyMoved`, `seldomMoved`, `oftenMoved`, and `alwaysMoved`, respectively; *OBJ*.`stability` can hold one value of the relative stabilities. The side of an object region is denoted by *OBJ*.`side`. In particular, *OBJ*.`front`, *OBJ*.`left`, *OBJ*.`back`, *OBJ*.`right`, *OBJ*.`upper`, and *OBJ*.`bottom` represent qualitative sides[1] of *OBJ*. *OBJ*.`side.stability` represents the relative stability of a side of *OBJ*, and *OBJ*.`side.stability`=*OBJ*.`stability`. The category of a side of an object region *OBJ*.`side` is written as `OBJ.side`, e.g., the category of *OBJ*.`front` is `OBJ.front`.

5.2 Spatial Relations between Regions

The distance relation and the orientation relation between object regions can be formalized using the connectedness relation.

5.2.1 'Connectedness' Is Primitive

A *region* refers to an *object region*, a *side region*, or a *constructed region* (defined later). The only primitive relation between regions is "connected with": **C**. In the literature, *that two regions are connected* is interpreted as *their closures share a point*[2] *in common*, i.e. Randell et al. (1992). When two regions are connected, for any category there is an object region in this category such that it is connected with both of the two regions. This is equivalent to that if two regions are not connected, then there is a category such that all regions in this category are not connected with both of the two objects. Let the category be the length unit which is smaller than the minimal distance between two disconnected regions, then any region in this category is not connected with both of the two disconnected regions. An axiom that governs the connectedness relation is as follows.

Axiom 5.2.1. *All regions A and B, it holds that if one is connected with the other, then all category* $\mathtt{Z}$ *there is a region Z satisfying that Z is a member of* $\mathtt{Z}$ *such that Z is connected with both A and B.*

$$\forall A,B \bullet \mathbf{C}(A,B) \rightarrow \forall \mathtt{Z} \exists Z | Z \in \mathtt{Z} \bullet \mathbf{C}(A,Z) \wedge \mathbf{C}(B,Z)$$

Another axioms governing the connectedness relation, following Randell et al. (1992), are as follows.

Axiom 5.2.2. *All region A, it holds that A is connected with A.*

$$\forall A \bullet \mathbf{C}(A,A)$$

[1] The side can be named by saliency of sides of object regions (the intrinsic reference framework) or by a *fiat* projection (the deictic reference framework, or the absolute reference framework).

[2] A *point* can be mereotopologically defined, see Eschenbach (1994).

Axiom 5.2.3. *All regions A and B, it holds that if A is connected with B, then B is also connected with A.*

$$\forall A, B \bullet \mathbf{C}(A,B) \rightarrow \mathbf{C}(B,A)$$

The relation of parthood $\mathbf{P}(A,B)$ can be defined immediately.

Definition 5.2.1. *Given two regions A and B, 'A is part of B' is defined as all region Z, it holds that if Z is connected with A, then Z is connected with B.*

$$\mathbf{P}(A,B) \stackrel{\text{def}}{=} \forall Z \bullet \mathbf{C}(Z,A) \rightarrow \mathbf{C}(Z,B)$$

P is governed by the axiom as follows.

Axiom 5.2.4. *All regions A and B, it holds that if each one is the part of the other, then they are the same.*

$$\forall A, B \bullet \mathbf{P}(A,B) \wedge \mathbf{P}(B,A) \rightarrow A = B$$

5.2.2 The Representation of Spatial Extensions

When Mr. Bertel sits on the couch, he stretches out his arm to reach the cup on the tea-table and drinking tea. The relative space that Mr. Bertel can reach with the help of his arms is larger than the space of his body. It is a spatial extension of his body. The boundary of the extended space is such that for any object connecting with it, Mr. Bertel can stretch out his arm and be connected with this object. So, the extended space is the sum of all possible locations of Mr. Bertel's arms including the space of his body.

Formally, let A (category $\mathtt{A}$), X (category $\mathtt{X}$) be two object regions. The spatial extension of A by X is called "*near* extension of A by X" can be defined as the sum[3] of all possible regions of category $\mathtt{X}$ that are connected with the region A and is written as A^X. A is called "the anchor region" and X is called "the extension region", shown in Figure 5.1.

Definition 5.2.2. *Given two object regions A and X (category* $\mathtt{X}$*), the near extension of A by X is defined as the Y that all object region W, it holds that W is connected with Y, if and only if there is an object region V satisfying that V is of the same category as X and connected with A such that V is connected with W.*

$$A^X \stackrel{\text{def}}{=}$$
$$\iota Y(\forall W \bullet (\mathbf{C}(W,Y) \equiv \exists V | V \in \mathtt{X} \wedge X \in \mathtt{X} \wedge \mathbf{C}(A,V) \bullet \mathbf{C}(W,V)))$$

[3] The definition of sum is a deviation from the sum definition in (Smith, 1996, p.289): *The sum of φers can be defined as that entity y which is such that, given any entity w, w overlaps with y if and only if w overlaps with something that φs. That is:* $\sigma x(\varphi x) := \iota y(\forall w(wOy \equiv \exists v(\varphi v \wedge wOv)))$.

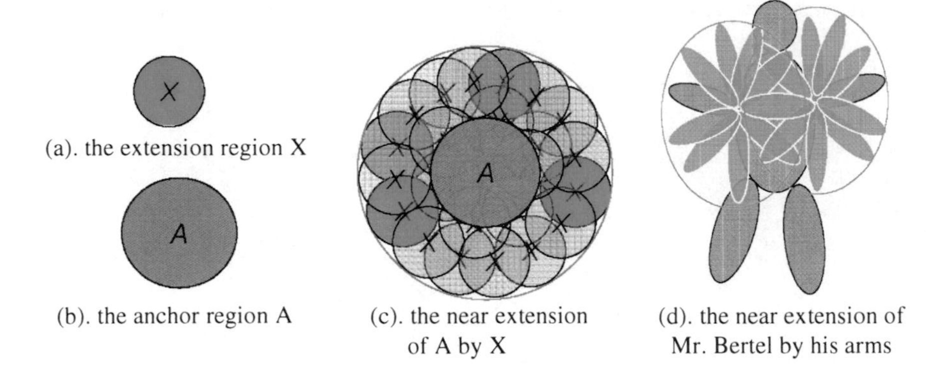

Fig. 5.1 (a) The extension region X; (b) the anchor region A; (c) the *near* extension region of A by X; (d) the near extension region of Mr. Bertel by his arms

The existence of A^X is guaranteed by **Theorem 5.2.1**; the uniqueness of A^X is guaranteed by **Axiom 5.2.4**. The near extension of A by X is called a *constructed region.* The category of A^X is written as '$\mathtt{A}^{\mathtt{X}}$'.

Theorem 5.2.1. *Given A an object region, and* $\mathtt{X}$ *be a category of an object region. There is X satisfying that X is a member of* $\mathtt{X}$ *such that if X is connected with A, then there is Y such that all W, it holds that Y is connected with W, if and only if there is V satisfying that V is a member of* $\mathtt{X}$ *and connected with A such that V is connected with W.*

$$\exists X | X \in \mathtt{X} \bullet \mathbf{C}(X,A) \rightarrow \exists Y \forall W \bullet (\mathbf{C}(W,Y) \equiv \exists V | V \in \mathtt{X} \wedge \mathbf{C}(A,V) \bullet \mathbf{C}(W,V)))$$

For example, let *BERTEL* be the object region of Mr. Bertel's body and *ARM* (category $\mathtt{ARM_{bio}}$) be the object region of his arm — the category $\mathtt{ARM_{bio}}$ has two features: (1) Elements are indistinguishable from one of Mr. Bertel's arms, (2) one end of each element is connected with one of Mr. Bertel's shoulder. Then the *near* extension of *BERTEL* by *ARM* ($BERTEL^{ARM}$) refers to the union of two object regions: (1) the object region of his body; and (2) the object region that his arm can reach, shown in Figure 5.1 (d).

Theorem 5.2.2. *All object regions A and X, it holds that A is a part of the near extension of A by X.*

$$\forall A, X \bullet \mathbf{P}(A, A^X)$$

Theorem 5.2.3. *All object regions A, B and X, it holds that the near extension of A by X is connected with B, if and only if A is connected with the near extension of B by X.*

$$\forall A, B, X \bullet \mathbf{C}(A^X, B) \equiv \mathbf{C}(A, B^X)$$

A constructed region can have its near extension region. For example, Mr. Bertel walks five steps from the couch to the table. By his first step, the near extension of the couch by his body is made; by his second step, the first near extension region is further extended with the condition that the body spaces of his first step and his second step are connected;.... The near extension of a constructed region by an object region is formalized as follows.

Definition 5.2.3. *Let C be a constructed region, and X be an object region (category* X*), the near extension of C by X is defined as the Y that all object region W, it holds that W is connected with Y, if and only if there is an object region V satisfying that V is of the same category as X and connected with C such that V is connected with W.*

$$C^X \stackrel{\text{def}}{=}$$
$$\iota Y(\forall W \bullet (\mathbf{C}(W,Y) \equiv \exists V | V \in \mathtt{X} \wedge X \in \mathtt{X} \wedge \mathbf{C}(C,V) \bullet \mathbf{C}(W,V)))$$

The existence of C^X is guaranteed by **Theorem 5.2.4**; the uniqueness of A^X is guaranteed by **Axiom 5.2.4**. The near extension of C by X is a *constructed region*. The category of C^X is written as '$\mathtt{C}^{\mathtt{X}}$'.

Theorem 5.2.4. *Let C be a constructed region, and* X *be a category of an object region. There is X satisfying that X is a member of* X *such that if X is connected with C, there is Y such that all W, it holds that Y is connected with W, if and only if there is V satisfying that V is a member of* X *and connected with C such that V is connected with W.*

$$\exists X | X \in \mathtt{X} \bullet \mathbf{C}(X,C) \rightarrow \exists Y \forall W \bullet (\mathbf{C}(W,Y) \equiv \exists V | V \in \mathtt{X} \wedge \mathbf{C}(C,V) \bullet \mathbf{C}(W,V)))$$

When Mr. Bertel sits on the couch, he can stretch out his arm and reach the cup of tea on the tea-table, however, he cannot reach the books on the writing-desk, no matter how he moves his two arms around. The books are located outside of the near extension region of his body by his arms. This results in the distance comparison that the books on the writing-desk are further away to him than the cup of tea on the tea-table. Formally, given an object region A, and object regions or side regions B and C, *that A is nearer to B than to C* can be defined as there is an extension region X such that the near extension of A by X is connected with B and disconnected with C.

Definition 5.2.4. *Let A be an object region, and B and C be object regions or side regions, then* that A is nearer to B than to C *is defined as there is an object region X, such that the near extension of A by X is connected with B and disconnected with C.*

$$\forall A,B,C \bullet nearer(A,B,C) \stackrel{\text{def}}{=} \exists X \bullet \mathbf{C}(A^X,B) \wedge \neg \mathbf{C}(A^X,C)$$

5.2.3 *Defining Qualitative Distances Using* Extension *Regions*

The distance between two disconnected objects can be defined by the amount of extension objects that are connected with both of them. For example, when

Mr. Bertel sits on the couch, it will take him seven steps to take the books on the writing-desk. The distance between Mr. Bertel sitting on the couch and the books on the writing-desk is "seven steps".

Formally, the distance from object region A to object region B is represented by the minimal number of extension regions (all are of category $\mathtt{X}$) such that the near extension of A by these regions is connected with B. If the first *near* extension region is named as "1X", the second *near* extension region is named as "2X", the third *near* extension region is named as "3X", ..., then a **naïve natural number system** for distance relation is created. Given two object regions A and B, "the distance from A to B is $n\mathtt{X}$" is interpreted as that B is connected with the nth *near* extension region X of A and disconnected with its $(n-1)^{th}$ *near* extension region. The qualitative distance from A to B is, therefore, defined in the notion of the naïve natural number system as follows.

Definition 5.2.5. *Let A, B be two object regions and $X_1, X_2, \ldots, X_n$ be n extension regions of category* $\mathtt{X}$, *the distance from A to B is defined as '$n\mathtt{X}$', if*

$$\neg\mathbf{C}((((A^{X_1})^{X_2})^{\cdots})^{X_{n-1}}, B) \wedge \mathbf{C}((((A^{X_1})^{X_2})^{\cdots})^{X_n}, B)$$

5.2.4 *Defining Qualitative Orientations Using the* Nearer *Predicate*

The orientation relation between two extended objects can be interpreted as the distance comparison between one extended object and the sides of the other object. For example, *that the balloon is in front of the writing-desk* can be interpreted as the balloon is nearer to the front side of the writing-desk than to its other sides. Formally, the orientation relation is formalized by the connectedness relation between a near extension of one object region and side regions of the other object region.

Definition 5.2.6. *Let A and B be two object regions, and B*.left, *B*.front, *B*.right, *B*.back *be four side regions of B. 'A is in front of B', written as* **Front**(A,B), *is defined as A is nearer to B*.front *than to B*.left, *B*.right, *and B*.back.

$$\mathbf{Front}(A,B) \stackrel{\text{def}}{=} \forall p | p \in \{B.\mathtt{left}, B.\mathtt{front}, B.\mathtt{right}, B.\mathtt{back}\}$$
$$\bullet p \neq B.\mathtt{front} \rightarrow nearer(A, B.\mathtt{front}, p)$$

Similarly, '**Left**(A,B)' stands for "A is left of B"; '**Right**(A,B)' for "A is right of B"; '**Behind**(A,B)' for "A is behind B".

Definition 5.2.7. *Let A and B be two object regions, and B*.left, *B*.front, *B*.right, *B*.back *be four side regions of B. 'A is left of B', written as* **Left**(A,B), *is defined as A is nearer to B*.left *than to B*.front, *B*.right, *and B*.back.

$$\mathbf{Left}(A,B) \stackrel{\text{def}}{=} \forall p | p \in \{B.\mathtt{left}, B.\mathtt{front}, B.\mathtt{right}, B.\mathtt{back}\}$$
$$\bullet p \neq B.\mathtt{left} \rightarrow nearer(A, B.\mathtt{left}, p)$$

Definition 5.2.8. *Let A and B be two object regions, and* $B.\texttt{left}$, $B.\texttt{front}$, $B.\texttt{right}$, $B.\texttt{back}$ *be four side regions of B. 'A is right of B', written as* $\textbf{Right}(A,B)$, *is defined as A is nearer to* $B.\texttt{right}$ *than to* $B.\texttt{left}$, $B.\texttt{front}$, *and* $B.\texttt{back}$.

$$\textbf{Right}(A,B) \stackrel{\text{def}}{=} \forall p | p \in \{B.\texttt{left}, B.\texttt{front}, B.\texttt{right}, B.\texttt{back}\} \\ \bullet p \neq B.\texttt{right} \rightarrow nearer(A, B.\texttt{right}, p)$$

Definition 5.2.9. *Let A and B be two object regions, and* $B.\texttt{left}$, $B.\texttt{front}$, $B.\texttt{right}$, $B.\texttt{back}$ *be four side regions of B. 'A is behind B', written as* $\textbf{Behind}(A,B)$, *is defined as A is nearer to* $B.\texttt{back}$ *than to* $B.\texttt{left}$, $B.\texttt{right}$, *and* $B.\texttt{front}$.

$$\textbf{Behind}(A,B) \stackrel{\text{def}}{=} \forall p | p \in \{B.\texttt{left}, B.\texttt{front}, B.\texttt{right}, B.\texttt{back}\} \\ \bullet p \neq B.\texttt{back} \rightarrow nearer(A, B.\texttt{back}, p)$$

Theorem 5.2.5. *Let A and B be two object regions,* $\textbf{Front}(A,B)$, $\textbf{Left}(A,B)$, $\textbf{Right}(A,B)$ *and* $\textbf{Behind}(A,B)$ *are pairwise disjoint.*

$$\begin{aligned} \textbf{Front}(A,B) \wedge \textbf{Left}(A,B) &\equiv \texttt{false} \\ \textbf{Left}(A,B) \wedge \textbf{Behind}(A,B) &\equiv \texttt{false} \\ \textbf{Behind}(A,B) \wedge \textbf{Right}(A,B) &\equiv \texttt{false} \\ \textbf{Right}(A,B) \wedge \textbf{Front}(A,B) &\equiv \texttt{false} \\ \textbf{Front}(A,B) \wedge \textbf{Behind}(A,B) &\equiv \texttt{false} \\ \textbf{Left}(A,B) \wedge \textbf{Right}(A,B) &\equiv \texttt{false} \end{aligned}$$

5.3 *Fiat* Containers: Formalizing Relative Spaces

Spatial relations between objects are interpreted as objects located in relative spaces in Chapter 4. For example, *that the cup is near Mr. Bertel* is interpreted as the cup is located in the relative space delineated by 'near Mr. Bertel'. This relative space can be defined by the region such that any object connected with it can be reached by Mr. Bertel's arm. Formally, this relative space *near Mr. Bertel* is called a **near *fiat* container**, *Mr. Bertel* is called the anchor of the *fiat* container, "be reached" can be formalized as "C". The near *fiat* container is formalized as follows.

Definition 5.3.1. *Let BERTEL be the object region of Mr. Bertel's body, ARM (category* $\texttt{ARM}_{\texttt{bio}}$*) be his arm, P be a region,* that *P* is near *BERTEL, written as* $\textbf{NR}_{ARM}(BERTEL,P)$, *is defined as that P is disconnected with BERTEL and connected with* $BERTEL^{ARM}$.

$$\textbf{NR}_{ARM}(BERTEL,P) \stackrel{\text{def}}{=} \neg\textbf{C}(BERTEL,P) \wedge \textbf{C}(BERTEL^{ARM},P)$$

Definition 5.3.2. *Let BERTEL be the object region of Mr. Bertel's body, ARM (category* $\mathtt{ARM_{bio}}$*) be his arm, then the* ***near fiat container****,* $\mathfrak{S}_{BERTEL}(\mathbf{NR}_{ARM})$*, is defined as the region Z that all region W, it holds that W is connected with Z, if and only if there is V satisfying that V is near BERTEL such that V is connected with W.*

$$\mathfrak{S}_{BERTEL}(\mathbf{NR}_{ARM})\stackrel{\text{def}}{=} \iota Z(\forall W \bullet (\mathbf{C}(W,Z) \equiv \exists V|\mathbf{NR}_{ARM}(BERTEL,V) \bullet \mathbf{C}(W,V)))$$

The existence of $\mathfrak{S}_{BERTEL}(\mathbf{NR}_{ARM})$ is guaranteed by **Theorem 5.3.1**; the uniqueness of $\mathfrak{S}_{BERTEL}(\mathbf{NR}_{ARM})$ is guaranteed by **Axiom 5.2.4**.

Theorem 5.3.1. *Given A an object region. There is X such that if X is near A, then there is Y such that all W, it holds that W is connected with Y, if and only if there is V satisfying that V is near A such that W is connected with V.*

$$\exists X \bullet \mathbf{NR}_{ARM}(A,X) \rightarrow \exists Y \forall W \bullet (\mathbf{C}(W,Y) \equiv \exists V|\mathbf{NR}_{ARM}(A,V) \bullet \mathbf{C}(W,V)))$$

In general, a ***fiat* container** of an object region is defined as follows.

Definition 5.3.3. *Let A be an object region, and* **rel** *be a spatial property (related with A), then the* ***rel* fiat *container****, written as* $\mathfrak{S}_A(\mathbf{rel})$*, is defined as the region Z that all region W, it holds that W is connected with Z, if and only if there is V satisfying the property* **rel** *such that V is connected*[4] *with W.*

$$\mathfrak{S}_A(\mathbf{rel})\stackrel{\text{def}}{=}\iota Z(\forall W \bullet (\mathbf{C}(W,Z) \equiv \exists V|\mathbf{rel}(V) \bullet \mathbf{C}(V,W)))$$

The existence of $\mathfrak{S}_A(\mathbf{rel})$ is guaranteed by **Theorem 5.3.2**; the uniqueness of $\mathfrak{S}_A(\mathbf{rel})$ is guaranteed by **Axiom 5.2.4**. $\mathfrak{S}_A(\mathbf{rel})$ is a *constructed region*. Its category is written as '$\mathtt{rel} * \mathtt{A}$'. *That a region P is connected with* $\mathfrak{S}_A(\mathbf{rel})$ is read as "*P* $\mathtt{rel}$ *A*".

Theorem 5.3.2. *Let* **rel** *be a spatial property. There is X such that if X satisfies* **rel***, then there is Y such that all W, it holds that W is connected with Y, if and only if there is V satisfying* **rel** *such that W is connected with V.*

$$\exists X \bullet \mathbf{rel}(X) \rightarrow \exists Y \forall W \bullet (\mathbf{C}(W,Y) \equiv \exists V|\mathbf{rel}(V) \bullet \mathbf{C}(W,V)))$$

Definition 5.3.4. *Let* $\mathfrak{S}_A(\mathbf{rel}_A)$ *and* $\mathfrak{S}_B(\mathbf{rel}_B)$ *be two* fiat *containers, V be an object region such that* $\mathbf{rel}_A(V)$ *and* $\mathbf{rel}_B(V)$*, then the* conjunction *of* $\mathfrak{S}_A(\mathbf{rel}_A)$ *and* $\mathfrak{S}_B(\mathbf{rel}_B)$*, written as* $\mathfrak{S}_A(\mathbf{rel}_A) \cap \mathfrak{S}_B(\mathbf{rel}_B)$*, is defined as the region Z that all W, it holds that W is connected with Z, if and only if there is V satisfying the property* $\mathbf{rel}_A$ *and* $\mathbf{rel}_B$ *such that V is connected with W.*

[4] *The connectedness relations between W and Z, and between V and the W here can be further generalized into a dyadic relation based on the connectedness relation,* $f(\mathbf{C})$*. A* fiat *container can, therefore, further generalized as:* $\mathfrak{S}_A(\mathbf{rel})\stackrel{\text{def}}{=}\iota Z(\forall W \bullet (f(\mathbf{C})(W,Z) \equiv \exists V|\mathbf{rel}(V) \bullet f(\mathbf{C})(V,W)))$.

$$\mathfrak{S}_A(\mathbf{rel}_A) \cap \mathfrak{S}_B(\mathbf{rel}_B) \stackrel{\text{def}}{=} \iota Z(\forall W \bullet (\mathbf{C}(W,Z) \equiv \exists V | \mathbf{rel}_A(V) \wedge \mathbf{rel}_B(V) \bullet \mathbf{C}(V,W)))$$

The existence of $\mathfrak{S}_A(\mathbf{rel}_A) \cap \mathfrak{S}_B(\mathbf{rel}_B)$ is guaranteed by **Theorem 5.3.3**; the uniqueness of $\mathfrak{S}_A(\mathbf{rel}_A) \cap \mathfrak{S}_B(\mathbf{rel}_B)$ is guaranteed by **Axiom 5.2.4**. Its category is written as '$\texttt{rel}_A * \texttt{A} + \texttt{rel}_B * \texttt{B}$'. *That a region P is connected with* $\mathfrak{S}_A(\mathbf{rel}_A) \cap \mathfrak{S}_B(\mathbf{rel}_B)$ is read as "P $\texttt{rel}_A$ A and $\texttt{rel}_B$ B".

Theorem 5.3.3. *Let* $\mathbf{rel}_A$ *and* $\mathbf{rel}_B$ *be two spatial properties that are related with regions A and B, respectively. There is a region V such that if V satisfies* $\mathbf{rel}_A(V)$ *and* $\mathbf{rel}_B(V)$, *then there is Y such that all W, it holds that W is connected with Y, if and only if there is V satisfying* $\mathbf{rel}_A$ *and* $\mathbf{rel}_B$ *such that W is connected with V.*

$$\exists X \bullet \mathbf{rel}_A(X) \wedge \mathbf{rel}_B(X) \rightarrow \exists Y \forall W \bullet (\mathbf{C}(W,Y) \equiv \exists V | \mathbf{rel}_A(V) \wedge \mathbf{rel}_B(V) \bullet \mathbf{C}(W,V)))$$

Let "`.anchor`" and "`.relation`" be two operators of a *fiat* container that return the anchor region and the relation of the *fiat* container, respectively. For example, $\mathfrak{S}_{COUCH}(\mathbf{NR}_{ARM})$`.anchor` $= COUCH$ and $\mathfrak{S}_{COUCH}(\mathbf{NR}_{ARM})$`.relation` $= \mathbf{NR}_{ARM}$; and let "`.objects`" be the operator of a *fiat* container that returns the set of object regions such that elements in the set are connected with this *fiat* container. These elements are called "located" in the *fiat* containers. For example, *that the tea-table is near the couch* is formally represented as '$TEATABLE \in \mathfrak{S}_{COUCH}(\mathbf{NR}_{ARM})$`.objects`'.

5.3.0.1 Connectedness *fiat* Containers

That an object region P is located in a connectedness fiat *container of an object region O*, $\mathfrak{S}_O(\mathbf{C})$, represents that P is connected with O. This depicts a trivial distance relation between P and O – they are connected. The category of a connectedness *fiat* container $\mathfrak{S}_O(\mathbf{C})$ is written as $\texttt{C} * \texttt{O}$, where `C` represents the connectedness relation, and `O` represents the category of the anchor object. For example, let *COUCH* and *TEATABLE* be object regions representing Mr. Bertel's couch and his tea-table, *ROOM* be the object region representing Mr. Bertel' room, then *that the couch and the tea-table are in Mr. Bertel's room* is formalized as

$$\mathfrak{S}_{ROOM}(\mathbf{C}).\texttt{objects} = \{COUCH, TEATABLE\}$$

The category of $\mathfrak{S}_{ROOM}(\mathbf{C})$ is $\texttt{C} * \texttt{ROOM}$. The sides of the room are used as objects, rather than object sides. *That the couch and the tea-table stand on the floor* is formalized as follows.

$$\mathfrak{S}_{FLOOR}(\mathbf{C}).\texttt{objects} = \{COUCH, TEATABLE\}$$

The category of $\mathfrak{S}_{FLOOR}(\mathbf{C})$ is $\texttt{C} * \texttt{FLOOR}$.

5.3.0.2 Distance *fiat* Containers

For disconnected object regions, their spatial relations can be specified by distance relations. Formally, one object region near the other object region can be defined in the notion of *fiat* containers as follows: For two disconnected object regions A and B, if B is near A, then B is located in the *near fiat* container of A. Let X be the extension region. This near relation $\mathbf{NR}_X$ is defined as follows.

$$\mathbf{NR}_X(A,B) \stackrel{\text{def}}{=} \neg\mathbf{C}(A,B) \wedge \mathbf{C}(A^X,B)$$

That an object region P is located in a near fiat *container of an object region O*, $\mathfrak{S}_O(\mathbf{NR}_X)$, represents that P is disconnected with O and connected with O^X. This depicts a distance relation between P and O – P is near O. The category of the *near fiat* container $\mathfrak{S}_O(\mathbf{NR}_X)$ is written as $\mathtt{NR_X} * \mathtt{O}$, where $\mathtt{NR_X}$ represents the near relation, and $\mathtt{O}$ represents the category of the anchor object. For example, Mr. Bertel's mother observed that the balloon was near the writing-desk and the tea-table was near the couch, as she could reach the writing-desk while sitting on the balloon and she could reach the tea-table while sitting on the couch (The extension object was her body, $BODY_m$.). Let *WRITINGDESK*, *COUCH*, *BALLOON* and *TEATABLE* be the object regions representing the writing-desk, the couch, the balloon, and the tea-table, respectively, the distance relations are formalized as follows.

$$\mathfrak{S}_{WRITINGDESK}(\mathbf{NR}_{BODY_m}).\mathtt{objects} = \{BALLOON\}$$
$$\mathfrak{S}_{COUCH}(\mathbf{NR}_{BODY_m}).\mathtt{objects} = \{TEATABLE\}$$

The category of $\mathfrak{S}_{WRITINGDESK}(\mathbf{NR}_{BODY_m})$ is $\mathtt{NR_{BODY_m}} * \mathtt{WRITINGDESK}$; and the category of $\mathfrak{S}_{COUCH}(\mathbf{NR}_{BODY_m})$ is $\mathtt{NR_{BODY_m}} * \mathtt{COUCH}$. *That the couch is in the corner* can be interpreted as the couch is near the two walls. Let *FRONTWALL* and *RIGHTWALL* be the front wall and the right wall. The location of the couch is formalized as follows.

$$\mathfrak{S}_{FRONTWALL}(\mathbf{NR}_{BODY_m}).\mathtt{objects} = \{COUCH\}$$
$$\mathfrak{S}_{RIGHTWALL}(\mathbf{NR}_{BODY_m}).\mathtt{objects} = \{COUCH\}$$

5.3.0.3 Orientation *fiat* Containers

Orientation relations are relations on distance comparison between an object region and sides of the other object region, for example, *that object region A is in front of object region B* means that A is nearer to $B.\mathtt{front}$ than to other sides of B ($B.\mathtt{left}$, $B.\mathtt{right}$, and $B.\mathtt{back}$). Formally, it is written as

$$\forall p | p \in \{B.\mathtt{front}, B.\mathtt{left}, B.\mathtt{right}, B.\mathtt{back}\}$$
$$\bullet p \neq B.\mathtt{front} \rightarrow nearer(A, B.\mathtt{front}, p)$$

This is defined as '$\mathbf{Front}(A,B)$' in 5.2.4. (page 60). The region Z that is delineated by 'in front of B' is such that all object region W, it holds that W is connected with

Z, if and only if there is an object region V satisfying that it is located 'in front of B' such that V is connected with W. The region Z is formalized as follows.

$$\mathfrak{S}_B(\mathbf{Front}) \stackrel{\text{def}}{=} \iota Z(\forall W \bullet (\mathbf{C}(W,Z) \equiv \exists V | \mathbf{Front}(V,B) \bullet \mathbf{C}(V,W)))$$

The category of $\mathfrak{S}_B(\mathbf{Front})$ is `Front` $*$ `B`. In general, the category of an orientation *fiat* container $\mathfrak{S}_O(\mathbf{Ori})$ is written as `Ori` $*$ `O`, where `Ori` represents an orientation relation, and `O` represents the category of the anchor object. For example, *that the balloon is in front of the writing-desk* and *that the tea-table is in front of the couch* are formalized as follows.

$$\mathfrak{S}_{WRITINGDESK}(\mathbf{Front}).\texttt{objects} = \{BALLOON\}$$
$$\mathfrak{S}_{COUCH}(\mathbf{Front}).\texttt{objects} = \{TEATABLE\}$$

The category $\mathfrak{S}_{WRITINGDESK}(\mathbf{Front})$ is written as `Front` $*$ `WRITINGDESK`; the category $\mathfrak{S}_{COUCH}(\mathbf{Front})$ is written as `Front` $*$ `COUCH`.

5.4 The Principle of Selecting *fiat* Containers

The principle of selecting cognitive reference objects can be stated in the notion of *fiat* container and called the principle of selecting *fiat* containers as follows: Let A, B_i ($\forall i \bullet 1 \le i \le n$) be object regions, A is referenced to connectedness, distance, or orientation *fiat* containers of $B_1, B_2, \ldots, B_n$, then (1) $A.\texttt{stability} < B_i.\texttt{stability}$ ($\forall i \bullet 1 \le i \le n$) and (2) for any object region C such that it is relatively more stable than region A, it holds that $nearer(A,B_i,C)$ ($\forall i \bullet 1 \le i \le n$). The first statement corresponds to the criterion of stability; the second statement corresponds to the criterion of economics.

5.5 Formalization of Cognitive Spectrums: $\mathfrak{C}$

A cognitive spectrum of indoor spatial environment can be symbolically represented by a table of objects and a table of relative spaces. The table of objects can be formalized by property operators of object regions, such as `.category`, `.side`, `.stability`, etc. Relative spaces are formalized by *fiat* containers. A cognitive spectrum $\mathfrak{C}$ of an indoor spatial environment can be formalized by a pair

$$\mathfrak{C} = < \langle \mathfrak{S} \rangle, ROOM >$$

$\langle \mathfrak{S} \rangle$ represents set of *fiat* containers and $ROOM$ is the object region of the room of this indoor environment. For example, a cognitive spectrum of the indoor spatial environment $\mathfrak{C}_{F4.7}$, shown in Figure 4.7, can be formalized as follows.

$$\mathfrak{C}_{F4.7} = <\langle\mathfrak{S}\rangle, ROOM>$$
$$\langle\mathfrak{S}\rangle = \{\mathfrak{S}_{ROOM}(\mathbf{C}), \mathfrak{S}_{FLOOR}(\mathbf{C}), \mathfrak{S}_{FRONTWALL}(\mathbf{NR}_{BODY_m})$$
$$\mathfrak{S}_{RIGHTWALL}(\mathbf{NR}_{BODY_m}), \mathfrak{S}_{COUCH}(\mathbf{NR}_{BODY_m}), \mathfrak{S}_{COUCH}(\mathbf{Front})\}$$
$$\mathfrak{S}_{ROOM}(\mathbf{C}).\texttt{objects} = \{COUCH, TEATABLE\}$$
$$\mathfrak{S}_{FLOOR}(\mathbf{C}).\texttt{objects} = \{COUCH, TEATABLE\}$$
$$\mathfrak{S}_{FRONTWALL}(\mathbf{NR}_{BODY_m}).\texttt{objects} = \{COUCH\}$$
$$\mathfrak{S}_{RIGHTWALL}(\mathbf{NR}_{BODY_m}).\texttt{objects} = \{COUCH\}$$
$$\mathfrak{S}_{COUCH}(\mathbf{NR}_{BODY_m}).\texttt{objects} = \{TEATABLE\}$$
$$\mathfrak{S}_{COUCH}(\mathbf{Front}).\texttt{objects} = \{TEATABLE\}$$

5.6 A Location and the Location

Definition 5.6.1. *Let A be an object region, and $\mathfrak{C}$ be a cognitive spectrum, $\mathfrak{S}$ be a* fiat *container of $\mathfrak{C}$, then $\mathfrak{S}$ is* a location *of A in $\mathfrak{C}$, if $A \in \mathfrak{S}.\texttt{objects}$.* The location *of A in $\mathfrak{C}$ is the conjunction of all locations of A in $\mathfrak{C}$, written as '$A.\texttt{Loc}|_{\mathfrak{C}}$' or '$A.\texttt{Loc}$' for short, if $\mathfrak{C}$ is clear.*

For example, Mr. Bertel's balloon is located on the floor, near the writing-desk and in front of the writing-desk. Let $BALLOON_B$, $WRITINGDESK_B$ and $FLOOR_B$ be object regions representing Mr. Bertel's balloon, his writing-desk, and the floor of his room, respectively, then *the location* of the balloon is formalized as follows.

$$\begin{aligned} BALLOON_B.\texttt{Loc} = & \; WRITINGDESK_B(\mathbf{Front}) \\ & \cap WRITINGDESK_B(\mathbf{NR}_{BODY_\mathbf{m}}) \\ & \cap FLOOR_B(\mathbf{C}) \end{aligned}$$

Suppose that there are n object regions in a cognitive spectrum $\mathfrak{C}$, then there will be $C*n$ *fiat* containers in $\mathfrak{C}$, where C is a constant; and the computational complexity of .Loc would be no higher than $O(n)$.

Definition 5.6.2. *Let $\mathfrak{S}_1, \ldots, \mathfrak{S}_n$ be n* fiat *containers, $\mathfrak{S} = \mathfrak{S}_1 \cap \cdots \cap \mathfrak{S}_n$, written as '$\bigcap_{i=1}^{n} \mathfrak{S}_i$', then $\mathfrak{S}.\texttt{Components} = \{\mathfrak{S}_1, \ldots, \mathfrak{S}_n\}$*

For example,

$$\begin{aligned} BALLOON_B.\texttt{Loc.Components} = & \; \{WRITINGDESK_B(\mathbf{Front}), \\ & WRITINGDESK_B(\mathbf{NR}_{BODY_\mathbf{m}}), \\ & FLOOR_B(\mathbf{C})\} \end{aligned}$$

Each component of the conjunction is *a location* of the balloon.

5.7 Relations between Two $\mathfrak{C}$s

5.7.1 The Primitive Relation

The primitive relation between two cognitive spectrums is the *categorically the same* relation between two object regions. Other relations are defined based on this primitive relation.

Definition 5.7.1. *Let OBJ_1 and OBJ_2 be two object regions, then OBJ_1 and OBJ_2 are* categorically the same, *written as $OBJ_1 \stackrel{\text{cat}}{=} OBJ_2$, if and only if*

$$OBJ_1.\texttt{category} = OBJ_2.\texttt{category}$$

$OBJ_1 \stackrel{\text{cat}}{\neq} OBJ_2$ denotes OBJ_1 and OBJ_2 are not categorically the same.

For example, Let $ROOM_B$, $BOOKSHELF_B$ be object regions representing Mr. Bertel's room and his bookshelf, and $ROOM_C$, $BOOKSHELF_C$ be object regions representing Mr. Certel's room and his bookshelf, then Mr. Bertel's bookshelf and Mr. Certel's bookshelf are *categorically the same*, $BOOKSHELF_B \stackrel{\text{cat}}{=} BOOKSHELF_C$. Mr. Certel's room $ROOM_C$ and Mr. Bertel's room $ROOM_B$ are not *categorically the same*, $ROOM_C \stackrel{\text{cat}}{\neq} ROOM_B$.

5.7.2 The Relations between Two Sets of Object Regions

Definition 5.7.2. *Let S be a set of object regions and OBJ be an object region, if there is an object region $O \in S$ such that $O \stackrel{\text{cat}}{=} OBJ$, then that S* categorically minuses *OBJ, written as '$S \ominus OBJ$', is defined as $S - \{O\}$.*

$$S \ominus OBJ \stackrel{\text{def}}{=} \begin{cases} S - \{O\} & \textit{if } \exists O | O \in S \bullet O \stackrel{\text{cat}}{=} OBJ \\ S & \textit{otherwise} \end{cases}$$

Suppose that S has n object regions, then the computational complexity of $S \ominus OBJ$ would be no higher than $O(n)$.

For example, let $ROOM_B$, $BOOKSHELF_B$ and $WRITINGDESK_B$ be Mr. Bertel's room, his bookshelf, and his writing-desk, and $BOOKSHELF_C$ be Mr. Certel's bookshelf, $S = \{ROOM_B, BOOKSHELF_B, WRITINGDESK_B\}$, then $S \ominus BOOKSHELF_C = \{ROOM_B, WRITINGDESK_B\}$.

Definition 5.7.3. *Let S_1 and S_2 be two sets of object regions, then S_1* categorically minuses *S_2, written as '$S_1 \ominus S_2$', is defined recursively as follows.*

$$S_1 \ominus S_2 \stackrel{\text{def}}{=} \begin{cases} S_1 & \textit{if } S_2 = \emptyset \\ (S_1 \ominus O') \ominus (S_2 - \{O'\}) & \exists O' \bullet O' \in S_2 \end{cases}$$

Suppose that each of S_1 and S_2 has at most n object regions, then the computational complexity of $S_1 \ominus S_2$ would be no higher than $O(n^2)$.

Definition 5.7.4. *Let S_1 and S_2 be two sets of object regions, then S_1 and S_2 are* categorically the same, *written as '$S_1 \stackrel{\text{cat}}{=} S_2$' if and only if*

$$S_1 \ominus S_2 = S_2 \ominus S_1 = \emptyset$$

For example, let $BOOKSHELF_B$ and $WRITINGDESK_B$ be Mr. Bertel's bookshelf and his writing-desk, $BOOKSHELF_C$ and $WRITINGDESK_C$ be Mr. Certel's bookshelf and his writing-desk, $S_B = \{BOOKSHELF_B, WRITINGDESK_B\}$ and $S_C = \{BOOKSHELF_C, WRITINGDESK_C\}$, then

$$\begin{aligned}
S_B \ominus S_C &= \{BOOKSHELF_B, WRITINGDESK_B\} \\
&\quad \ominus \{BOOKSHELF_C, WRITINGDESK_C\} \\
&= (\{BOOKSHELF_B, WRITINGDESK_B\} \ominus BOOKSHELF_C) \\
&\quad \ominus (\{BOOKSHELF_C, WRITINGDESK_C\} - \{BOOKSHELF_C\}) \\
&= \{WRITINGDESK_B\} \ominus \{WRITINGDESK_C\} \\
&= \emptyset \\
S_C \ominus S_B &= \{BOOKSHELF_C, WRITINGDESK_C\} \\
&\quad \ominus \{BOOKSHELF_B, WRITINGDESK_B\} \\
&= (\{BOOKSHELF_C, WRITINGDESK_C\} \ominus BOOKSHELF_B) \\
&\quad \ominus (\{BOOKSHELF_B, WRITINGDESK_B\} - \{BOOKSHELF_B\}) \\
&= \{WRITINGDESK_C\} \ominus \{WRITINGDESK_B\} \\
&= \emptyset
\end{aligned}$$

Therefore, S_B and S_C are *categorically the same*: $S_B \stackrel{\text{cat}}{=} S_C$.

Note that the result of $\ominus$ is unique with respect to the *categorically the same* relation: $\stackrel{\text{cat}}{=}$, rather than with respect to the classic equal relation. For example, let S be a set of object regions, $S = \{BOOKSHELF_1, BOOKSHELF_2\}$, such that $BOOKSHELF_1$ and $BOOKSHELF_2$ are of the same category, then the result of $S \ominus \{BOOKSHELF_1\}$ can be $R_1 = \{BOOKSHELF_1\}$ or $R_2 = \{BOOKSHELF_2\}$. It holds that $R_1 \stackrel{\text{cat}}{=} R_2$, but not $R_1 = R_2$.

5.7.3 The Relation between fiat *Containers*

Definition 5.7.5. *Let $\mathfrak{S}_1$ and $\mathfrak{S}_2$ be two* fiat *containers, if they are of the same category, then they are* categorically the same, *written as '$\mathfrak{S}_1 \stackrel{\text{cat}}{=} \mathfrak{S}_2$'.*

For example, let $WRITINGDESK_B$ and $WRITINGDESK_C$ be Mr. Bertel's writing-desk and Mr. Certel's writing-desk, respectively, the category of $\mathfrak{S}_{WRITINGDESK_B}(\mathbf{C})$ is $\texttt{C} * \texttt{WRITINGDESK}$, the category of $\mathfrak{S}_{WRITINGDESK_C}(\mathbf{C})$ is $\texttt{C} * \texttt{WRITINGDESK}$, therefore, $\mathfrak{S}_{WRITINGDESK_B}(\mathbf{C})$ and $\mathfrak{S}_{WRITINGDESK_C}(\mathbf{C})$ are *categorically the same*. The category of $\mathfrak{S}_{WRITINGDESK_B}(\mathbf{NR}_{BODY_m})$ is $\texttt{NR}_{\texttt{BODY}_\texttt{m}} * \texttt{WRITINGDESK}$, the category of $\mathfrak{S}_{WRITINGDESK_C}(\mathbf{NR}_{BODY_m})$ is $\texttt{NR}_{\texttt{BODY}_\texttt{m}} * \texttt{WRITINGDESK}$, therefore, the distance *fiat* containers $\mathfrak{S}_{WRITINGDESK_B}(\mathbf{NR}_{BODY_m})$ and $\mathfrak{S}_{WRITINGDESK_C}(\mathbf{NR}_{BODY_m})$ are *categorically the same*. The category of $\mathfrak{S}_{WRITINGDESK_B}(\mathbf{Front})$ is `Front` $*$ `WRITINGDESK`, the category of $\mathfrak{S}_{WRITINGDESK_C}(\mathbf{Front})$ is `Front` $*$ `WRITINGDESK`, therefore, the orientation *fiat* containers $\mathfrak{S}_{WRITINGDESK_B}(\mathbf{Front})$ and $\mathfrak{S}_{WRITINGDESK_C}(\mathbf{Front})$ are *categorically the same*.

5.7.4 The Relation between Two Sets of fiat *Containers*

Definition 5.7.6. *Let* $\langle\mathfrak{S}\rangle$ *be a set of* fiat *containers and* $\mathfrak{S}'$ *be a* fiat *container, if there is a* fiat *container* $\mathfrak{S}_1 \in \langle\mathfrak{S}\rangle$ *such that* $\mathfrak{S}_1 \overset{\text{cat}}{=} \mathfrak{S}'$, *then* $\langle\mathfrak{S}\rangle$ categorically minuses $\mathfrak{S}'$, *written as* '$\langle\mathfrak{S}\rangle \ominus \mathfrak{S}'$', *is defined as* $\langle\mathfrak{S}\rangle - \{\mathfrak{S}_1\}$

$$\langle\mathfrak{S}\rangle \ominus \mathfrak{S}' \overset{\text{def}}{=} \begin{cases} \langle\mathfrak{S}\rangle - \{\mathfrak{S}_1\} & \textit{if}\ \exists \mathfrak{S}_1 | \mathfrak{S}_1 \in \langle\mathfrak{S}\rangle \bullet \mathfrak{S}_1 \overset{\text{cat}}{=} \mathfrak{S}' \\ \langle\mathfrak{S}\rangle & \textit{otherwise} \end{cases}$$

Suppose that $\langle\mathfrak{S}\rangle$ has n *fiat* containers, then the computational complexity of $\langle\mathfrak{S}\rangle \ominus \mathfrak{S}$ would be no higher than $O(n)$.

Definition 5.7.7. *Let* $\langle\mathfrak{S}\rangle_1$ *and* $\langle\mathfrak{S}\rangle_2$ *be two sets of* fiat *containers, then* $\langle\mathfrak{S}\rangle_1$ categorically minuses $\langle\mathfrak{S}\rangle_2$, *written as* '$\langle\mathfrak{S}\rangle_1 \ominus \langle\mathfrak{S}\rangle_2$', *is defined recursively as follows*

$$\langle\mathfrak{S}\rangle_1 \ominus \langle\mathfrak{S}\rangle_2 \overset{\text{def}}{=} \begin{cases} \langle\mathfrak{S}\rangle_1 & \textit{if}\ \langle\mathfrak{S}\rangle_2 = \emptyset \\ ((\langle\mathfrak{S}\rangle_1 \ominus \mathfrak{S}') \ominus (\langle\mathfrak{S}\rangle_2 - \{\mathfrak{S}'\}) & \exists \mathfrak{S}' \bullet \mathfrak{S}' \in \langle\mathfrak{S}\rangle_2 \end{cases}$$

Suppose that each of $\langle\mathfrak{S}\rangle_1$ and $\langle\mathfrak{S}\rangle_2$ has at most n *fiat* containers, then the computational complexity of $\langle\mathfrak{S}\rangle_1 \ominus \langle\mathfrak{S}\rangle_2$ would be no higher than $O(n^2)$.

Definition 5.7.8. *Let* $\langle\mathfrak{S}\rangle_1$ *and* $\langle\mathfrak{S}\rangle_2$ *be two sets of* fiat *containers, then* $\langle\mathfrak{S}\rangle_1$ *and* $\langle\mathfrak{S}\rangle_2$ *are* categorically the same, *written as* '$\langle\mathfrak{S}\rangle_1 \overset{\text{cat}}{=} \langle\mathfrak{S}\rangle_2$' *if and only if*

$$\langle\mathfrak{S}\rangle_1 \ominus \langle\mathfrak{S}\rangle_2 = \langle\mathfrak{S}\rangle_2 \ominus \langle\mathfrak{S}\rangle_1 = \emptyset$$

Definition 5.7.9. *Let* A, B *be two object regions, then their locations* A.`Loc` *and* B.`Loc` *are* categorically the same, *if* A.`Loc.Components`$\overset{\text{cat}}{=}B$.`Loc.Components`, *written as* 'A.`Loc`$\overset{\text{cat}}{=}B$.`Loc`'.

5.7.5 Mapping Object Regions and Mapping fiat *Containers*

Definition 5.7.10. *Let* $\mathfrak{C}_1$ *and* $\mathfrak{C}_2$ *be two cognitive spectrums,* $\mathfrak{C}_1 =< \langle\mathfrak{S}\rangle_1, ROOM_1 >$, $\mathfrak{C}_2 =< \langle\mathfrak{S}\rangle_2, ROOM_2 >$, $ROOM_1$ *and* $ROOM_2$ *are mapped,* '$ROOM_1 \overset{\text{map}}{=} ROOM_2$', *if and only if they are* categorically the same.

$$ROOM_1 \overset{\text{map}}{=} ROOM_2 \overset{\text{def}}{=} ROOM_1 \overset{\text{cat}}{=} ROOM_2$$

Definition 5.7.11. *Let $\langle\mathfrak{S}\rangle$ be a set of* fiat *containers, A be an object region, then* A.`fiatContainers` *is the subset of $\langle\mathfrak{S}\rangle$ such that the anchor object of each* fiat *container in the subset is A or one side of A.*

For example, let $\langle\mathfrak{S}\rangle$ be the set of *fiat* containers as follows.

$$\langle\mathfrak{S}\rangle = \{\mathfrak{S}_{ROOM_B}(\mathbf{C}), \mathfrak{S}_{FLOOR_B}(\mathbf{C}), \mathfrak{S}_{FRONTWALL_B}(\mathbf{NR}_{BODY_m}), \mathfrak{S}_{RIGHTWALL_B}(\mathbf{NR}_{BODY_m}), \mathfrak{S}_{COUCH_B}(\mathbf{Front})\}$$

Then, $ROOM_B$.`fiatContainers` of $\langle\mathfrak{S}\rangle$ is the subset as follows.

$$ROOM_B.\texttt{fiatContainers} = \{\mathfrak{S}_{ROOM_B}(\mathbf{C}), \mathfrak{S}_{FLOOR_B}(\mathbf{C}), \mathfrak{S}_{FRONTWALL_B}(\mathbf{NR}_{BODY_m}), \mathfrak{S}_{RIGHTWALL_B}(\mathbf{NR}_{BODY_m})\}$$

Suppose that $\langle\mathfrak{S}\rangle$ has n elements, then the computational complexity of `.fiat–Containers` is no higher than $O(n)$.

Definition 5.7.12. *Two* fiat *containers $\mathfrak{S}_1$ and $\mathfrak{S}_2$ are* mapped, *written as '$\mathfrak{S}_1 \overset{\text{map}}{=} \mathfrak{S}_2$', if they are* categorically the same *and their anchor objects are* mapped.

$$\mathfrak{S}_1 \overset{\text{map}}{=} \mathfrak{S}_2 \overset{\text{def}}{=} (\mathfrak{S}_1 \overset{\text{cat}}{=} \mathfrak{S}_2) \wedge (\mathfrak{S}_1.\texttt{anchor} \overset{\text{map}}{=} \mathfrak{S}_2.\texttt{anchor})$$

Definition 5.7.13. *That two conjunctions of* fiat *containers $\bigcap_{i=1}^{n} \mathfrak{S}_{1i}$ and $\bigcap_{j=1}^{m} \mathfrak{S}_{2j}$ are* mapped, *written as '$\bigcap_{i=1}^{n} \mathfrak{S}_{1i} \overset{\text{map}}{=} \bigcap_{j=1}^{m} \mathfrak{S}_{2j}$', is recursively defined as follows.*

$$\bigcap_{i=1}^{n} \mathfrak{S}_{1i} \overset{\text{map}}{=} \bigcap_{j=1}^{m} \mathfrak{S}_{2j} \overset{\text{def}}{=} \begin{cases} \mathfrak{S}_{11} \overset{\text{map}}{=} \mathfrak{S}_{21} & n = m = 1 \\ (\mathfrak{S}_{11} \overset{\text{map}}{=} \mathfrak{S}_{2k}) \wedge (\bigcap_{i=2}^{n} \mathfrak{S}_{1i} \overset{\text{map}}{=} \bigcap_{j=1, j\neq k}^{m} \mathfrak{S}_{2j}) & n = m > 1 \\ \texttt{false} & n \neq m \end{cases}$$

Suppose that each of the conjunctions of *fiat* containers $\bigcap_{i=1}^{n} \mathfrak{S}_{1i}$ and $\bigcap_{j=1}^{m} \mathfrak{S}_{2j}$ has at most n components, then the computational complexity of $\bigcap_{i=1}^{n} \mathfrak{S}_{1i} \overset{\text{map}}{=} \bigcap_{j=1}^{m} \mathfrak{S}_{2j}$ is no higher than $O(n^2)$.

Definition 5.7.14. *Let $\mathfrak{C}_1$ and $\mathfrak{C}_2$ be two cognitive spectrums, $\mathfrak{C}_1 =< \langle\mathfrak{S}\rangle_1, ROOM_1 >$, $\mathfrak{C}_2 =< \langle\mathfrak{S}\rangle_2, ROOM_2 >$, A and B be two object regions in $\mathfrak{S}_{ROOM_1}(\mathbf{C})$ and $\mathfrak{S}_{ROOM_2}(\mathbf{C})$ respectively. A and B are mapped, written as '$A \overset{\text{map}}{=} B$', if and only if they are* categorically the same *and their locations in $\mathfrak{C}_1$ and $\mathfrak{C}_2$ are* mapped.

$$A \overset{\text{map}}{=} B \overset{\text{def}}{=} (A \overset{\text{cat}}{=} B) \wedge (A.\texttt{Loc}|_{\mathfrak{C}_1} \overset{\text{map}}{=} B.\texttt{Loc}|_{\mathfrak{C}_2})$$

Suppose that each of $\mathfrak{C}_1$ and $\mathfrak{C}_2$ has at most n object regions, then each of them has at most $C * n$ *fiat* containers, where C is a constant, the computational complexity of $A \overset{\text{map}}{=} B$ is no higher than the multiplication of the computational complexity of `.Loc`,

$O(n)$, and the computational complexity of $A.\texttt{Loc}|_{\mathfrak{C}_1} \stackrel{\text{map}}{=} B.\texttt{Loc}|_{\mathfrak{C}_2}$, $O(n^2)$. That is, the computational complexity of $A \stackrel{\text{map}}{=} B$ is no higher than $O(n^3)$.

At his mother's first and second visits, Mr. Bertel's couch is located in the corner. Let $ROOM_1$, $COUCH_1$, $FRONTWALL_1$, $RIGHTWALL_1$ and $FLOOR_1$ be object regions representing Mr. Bertel's room, the couch, the front wall, the right wall[5], and the floor at his mother's first visit; $ROOM_2$, $COUCH_2$, $FRONTWALL_2$, $RIGHTWALL_2$ and $FLOOR_2$ be object regions representing Mr. Bertel's room, the couch, the front wall, the right wall[6], and the floor at his mother's second visit, $\langle\mathfrak{S}\rangle_1$ and $\langle\mathfrak{S}\rangle_2$ be two sets of *fiat* containers defined as follows.

$$\begin{aligned}
\langle\mathfrak{S}\rangle_1 = \{&\mathfrak{S}_{ROOM_1}(\mathbf{C}), \mathfrak{S}_{FRONTWALL_1}(\mathbf{NR}_{BODY_m}),\\
&\mathfrak{S}_{RIGHTWALL_1}(\mathbf{NR}_{BODY_m}), \mathfrak{S}_{FLOOR_1}(\mathbf{C})\}\\
\langle\mathfrak{S}\rangle_2 = \{&\mathfrak{S}_{ROOM_2}(\mathbf{C}), \mathfrak{S}_{FRONTWALL_2}(\mathbf{NR}_{BODY_m}),\\
&\mathfrak{S}_{RIGHTWALL_2}(\mathbf{NR}_{BODY_m}), \mathfrak{S}_{FLOOR_2}(\mathbf{C})\}
\end{aligned}$$

The location of $COUCH_1$ and the location of $COUCH_2$ are

$$\begin{aligned}
COUCH_1.\texttt{Loc} = &\mathfrak{S}_{ROOM_1}(\mathbf{C}) \cap \mathfrak{S}_{FRONTWALL_1}(\mathbf{NR}_{BODY_m})\\
&\cap \mathfrak{S}_{RIGHTWALL_1}(\mathbf{NR}_{BODY_m}) \cap \mathfrak{S}_{FLOOR_1}(\mathbf{C})\\
COUCH_2.\texttt{Loc} = &\mathfrak{S}_{ROOM_2}(\mathbf{C}) \cap \mathfrak{S}_{FRONTWALL_2}(\mathbf{NR}_{BODY_m})\\
&\cap \mathfrak{S}_{RIGHTWALL_2}(\mathbf{NR}_{BODY_m}) \cap \mathfrak{S}_{FLOOR_2}(\mathbf{C})
\end{aligned}$$

Then, $COUCH_1$ and $COUCH_2$ are *mapped*, $COUCH_1 \stackrel{\text{map}}{=} COUCH_2$, if

$$\begin{aligned}
&(1)\ COUCH_1 \stackrel{\text{cat}}{=} COUCH_2\\
&(2)\ COUCH_1.\texttt{Loc} \stackrel{\text{map}}{=} COUCH_2.\texttt{Loc}
\end{aligned}$$

Definition 5.7.15. *Let $\langle\mathfrak{S}\rangle_1$ and $\langle\mathfrak{S}\rangle_2$ be two sets of* fiat *containers, the mapping between $\langle\mathfrak{S}\rangle_1$ and $\langle\mathfrak{S}\rangle_2$, written as '$\texttt{mapping}_{\mathfrak{f}}(\langle\mathfrak{S}\rangle_1, \langle\mathfrak{S}\rangle_2)$', results in a queue[7] of pairs such that two components of each pair are* mapped *and that they are members of $\langle\mathfrak{S}\rangle_1$ and $\langle\mathfrak{S}\rangle_2$, respectively.*

$$\texttt{mapping}_{\mathfrak{f}}(\langle\mathfrak{S}\rangle_1, \langle\mathfrak{S}\rangle_2) \stackrel{\text{def}}{=} [(\mathfrak{S}_{1i}, \mathfrak{S}_{2j}) | \mathfrak{S}_{1i} \in \langle\mathfrak{S}\rangle_1 \wedge \mathfrak{S}_{2j} \in \langle\mathfrak{S}\rangle_2 \wedge \mathfrak{S}_{1i} \stackrel{\text{map}}{=} \mathfrak{S}_{2j}]$$

Suppose that each of $\langle\mathfrak{S}\rangle_1$ and $\langle\mathfrak{S}\rangle_2$ has at most n *fiat* containers, then the computational complexity of $\texttt{mapping}_{\mathfrak{f}}(\langle\mathfrak{S}\rangle_1, \langle\mathfrak{S}\rangle_2)$ is no higher than $O(n^2)$.

[5] The front wall and the right wall are named by the same *fiat* projection as that in Figure 4.7, see note 7, on page 43.

[6] See the above note.

[7] A queue $\mathbb{A}$ is written as $[a_0, a_1, \ldots, a_n]$, $\mathbb{A}[i] = a_i$; that a is a component of $\mathbb{A}$ is written as $a \in \mathbb{A}$.

For example, let $COUCH_1$ and $COUCH_2$ be two object regions representing Mr. Bertel's couch of his mother's first and second visits, and $COUCH_1 \overset{\text{map}}{=} COUCH_2$; let $\langle\mathfrak{S}\rangle_1$ and $\langle\mathfrak{S}\rangle_2$ be sets of *fiat* containers as follows.

$$\langle\mathfrak{S}\rangle_1 = \{\mathfrak{S}_{COUCH_1}(\mathbf{Front}), \mathfrak{S}_{COUCH_1}(\mathbf{NR}_{BODY_m})\}$$
$$\langle\mathfrak{S}\rangle_2 = \{\mathfrak{S}_{COUCH_2}(\mathbf{Front}), \mathfrak{S}_{COUCH_2}(\mathbf{NR}_{BODY_m})\}$$

Then, $\langle\mathfrak{S}\rangle_1$ and $\langle\mathfrak{S}\rangle_2$ are mapped as follows.

$$\begin{aligned}\texttt{mapping}_\mathtt{f}(\langle\mathfrak{S}\rangle_1, \langle\mathfrak{S}\rangle_2) &= [(\mathfrak{S}_{COUCH_1}(\mathbf{Front}), \mathfrak{S}_{COUCH_2}(\mathbf{Front})),\\ &\qquad (\mathfrak{S}_{COUCH_1}(\mathbf{NR}_{BODY_m}), \mathfrak{S}_{COUCH_2}(\mathbf{NR}_{BODY_m}))]\\ \texttt{mapping}_\mathtt{f}(\langle\mathfrak{S}\rangle_1, \langle\mathfrak{S}\rangle_2)[0] &= (\mathfrak{S}_{COUCH_1}(\mathbf{Front}), \mathfrak{S}_{COUCH_2}(\mathbf{Front}))\\ \texttt{mapping}_\mathtt{f}(\langle\mathfrak{S}\rangle_1, \langle\mathfrak{S}\rangle_2)[1] &= (\mathfrak{S}_{COUCH_1}(\mathbf{NR}_{BODY_m}), \mathfrak{S}_{COUCH_2}(\mathbf{NR}_{BODY_m}))\end{aligned}$$

Definition 5.7.16. *Let* $\langle\mathfrak{S}\rangle_1$ *and* $\langle\mathfrak{S}\rangle_2$ *be two sets of* fiat *containers,* $\mathbb{Q}$ *be the result of* $\texttt{mapping}_\mathtt{f}(\langle\mathfrak{S}\rangle_1, \langle\mathfrak{S}\rangle_2)$*, then* $\mathbb{Q}.\texttt{first}$ *be the set of the first component of each pair in* $\mathbb{Q}$*,* $\mathbb{Q}.\texttt{second}$ *be the set of the second component of each pair in* $\mathbb{Q}$*.*

$$\mathbb{Q}.\texttt{first} = \{\mathfrak{S} | (\mathfrak{S}, \mathfrak{S}') \in \mathbb{Q}\}$$
$$\mathbb{Q}.\texttt{second} = \{\mathfrak{S}' | (\mathfrak{S}, \mathfrak{S}') \in \mathbb{Q}\}$$

For example,

$$\texttt{mapping}_\mathtt{f}(\langle\mathfrak{S}\rangle_1, \langle\mathfrak{S}\rangle_2).\texttt{first} = \{\mathfrak{S}_{COUCH_1}(\mathbf{Front}), \mathfrak{S}_{COUCH_1}(\mathbf{NR}_{BODY_m})\}$$
$$\texttt{mapping}_\mathtt{f}(\langle\mathfrak{S}\rangle_1, \langle\mathfrak{S}\rangle_2).\texttt{second} = \{\mathfrak{S}_{COUCH_2}(\mathbf{Front}), \mathfrak{S}_{COUCH_2}(\mathbf{NR}_{BODY_m})\}$$

Let (A,B) be a pair, then $(A,B).\texttt{first} = A$, $(A,B).\texttt{second} = B$. For example,

$$\texttt{mapping}_\mathtt{f}(\langle\mathfrak{S}\rangle_1, \langle\mathfrak{S}\rangle_2)[0].\texttt{first} = \mathfrak{S}_{COUCH_1}(\mathbf{Front})$$
$$\texttt{mapping}_\mathtt{f}(\langle\mathfrak{S}\rangle_1, \langle\mathfrak{S}\rangle_2)[0].\texttt{second} = \mathfrak{S}_{COUCH_1}(\mathbf{NR}_{BODY_m})$$

Definition 5.7.17. *Let* $\mathfrak{C}_1$ *and* $\mathfrak{C}_2$ *be two cognitive spectrums,* $\mathfrak{C}_1 =< \langle\mathfrak{S}\rangle_1, ROOM_1 >$, $\mathfrak{C}_2 =< \langle\mathfrak{S}\rangle_2, ROOM_2 >$, $\mathfrak{S}_1$ *and* $\mathfrak{S}_2$ *be two mapped* fiat *containers in the two cognitive spectrums, respectively. Objects in* $\mathfrak{S}_1$ *and* $\mathfrak{S}_2$ *are* mapped*, if they are* categorically the same *and their locations in* $\mathfrak{C}_1$ *and* $\mathfrak{C}_2$ *are* mapped. *The object mapping process, written as '*$\texttt{mapping}_\mathtt{o}(\mathfrak{S}_1, \mathfrak{S}_2)$*', results in a queue of object region pairs such that two object regions in each pair are* mapped.

$$\texttt{mapping}_\mathtt{o}(\mathfrak{S}_1, \mathfrak{S}_2) \overset{\text{def}}{=} [(O_1, O_2) | O_1 \in \mathfrak{S}_1.\texttt{objects} \wedge O_2 \in \mathfrak{S}_2.\texttt{objects} \wedge O_1 \overset{\text{map}}{=} O_2]$$

Two objects cannot be *mapped*, before their locations are *mapped*. Their locations which are conjunctions of *fiat* containers cannot be *mapped*, before the anchor objects of the *fiat* containers are mapped. Let $\mathbb{Q}$ be the currently queue of the mapped *fiat* containers. The issue of whether two objects O_1 and O_2 are *mapped* can be addressed, if the components of their locations are subsets of $\mathbb{Q}$.`first` and $\mathbb{Q}$.`second`, respectively. That is,

$$O_1.\texttt{Loc}|_{\mathfrak{C}_1}.\texttt{Components} \subseteq \mathbb{Q}.\texttt{first}$$
$$O_2.\texttt{Loc}|_{\mathfrak{C}_2}.\texttt{Components} \subseteq \mathbb{Q}.\texttt{second}$$

The process of $\texttt{mapping}_o(\mathfrak{S}_1, \mathfrak{S}_2)$ is formalized as follows.

Input: Let $\mathfrak{C}_1$ and $\mathfrak{C}_2$ be two cognitive spectrums; let $\mathfrak{S}_1$ and $\mathfrak{S}_2$ be two *fiat* containers, and $\mathbb{Q}$ be the queue of the mapped *fiat* containers.
Output: Let $\mathbb{O}$ be the queue of mapped objects.
Initial state: $\mathbb{O} \leftarrow \emptyset$.
Process:

for each object region $\texttt{region}_{1i}$ in $\mathfrak{S}_1$.`objects` do
  for each object region $\texttt{region}_{2j}$ in $\mathfrak{S}_2$.`objects` do
    if $\texttt{region}_{1i} \stackrel{\text{cat}}{=} \texttt{region}_{2j}$ and
      $\texttt{region}_{1i}.\texttt{Loc}|_{\mathfrak{C}_1} \subseteq \mathbb{Q}.\texttt{first}$ and
      $\texttt{region}_{2j}.\texttt{Loc}|_{\mathfrak{C}_2} \subseteq \mathbb{Q}.\texttt{second}$ and
      $\texttt{region}_{1i}.\texttt{Loc}|_{\mathfrak{C}_1} \stackrel{\text{map}}{=} \texttt{region}_{2j}.\texttt{Loc}|_{\mathfrak{C}_2}$
    then append $(\texttt{region}_{1i}, \texttt{region}_{2j})$ to $\mathbb{O}$

Suppose that each of $\mathfrak{C}_1$ and $\mathfrak{C}_2$ has at most n object regions, then the computational complexity of $\texttt{mapping}_o(\mathfrak{S}_1, \mathfrak{S}_2)$ would be no higher than $O(n) * O(n) * (O(n^2) + O(n^2) + O(n^3)) = O(n^5)$.

5.7.5.1 Mapping Cognitive Spectrums

Input: Let $\mathfrak{C}_1$ and $\mathfrak{C}_2$ be two cognitive spectrums such that

$$\mathfrak{C}_1 = < \langle \mathfrak{S} \rangle_1, ROOM_1 >$$
$$\mathfrak{C}_2 = < \langle \mathfrak{S} \rangle_2, ROOM_2 >$$

Let $\mathbb{O}$ be the queue of mapped objects, $\mathbb{Q}$ be the queue of the mapped *fiat* containers, $\texttt{Index}_o$ be the index of the queue $\mathbb{O}$, and $\texttt{Index}_f$ be the index of the queue $\mathbb{Q}$.
Output: Let $\mathbb{O}$ be the queue of mapped objects.
Initial state: $\mathbb{O} \leftarrow \emptyset$, $\mathbb{Q} \leftarrow \emptyset$, $\texttt{Index}_f \leftarrow 0$, $\texttt{Index}_o \leftarrow 0$
Process:

(1) if $ROOM_1 \stackrel{\text{cat}}{=} ROOM_2$
  then $\mathbb{O} \leftarrow [(ROOM_1, ROOM_2)]$
    $\mathbb{Q} \leftarrow \texttt{mapping}_f(ROOM_1.\texttt{fiatContainers}, ROOM_2.\texttt{fiatContainers})$
  else go to (5)
(2) while $\texttt{Index}_f$ points to a pair in $\mathbb{Q}$, do
  (2.1) Append the result of $\texttt{mapping}_o\mathbb{Q}[\texttt{Index}_f]$ to $\mathbb{O}$

(2.2) $\mathtt{Index_f} \leftarrow \mathtt{Index_f} + 1$
(3) while $\mathtt{Index_o}$ points to a pair in $\mathbb{O}$, do
(3.1) $\langle\mathfrak{S}\rangle_1 \leftarrow \mathbb{O}[\mathtt{Index_o}]\mathtt{.first.fiatContainers}$
(3.2) $\langle\mathfrak{S}\rangle_2 \leftarrow \mathbb{O}[\mathtt{Index_o}]\mathtt{.second.fiatContainers}$
(3.3) Concatenate the result of $\mathtt{mapping_f}(\langle\mathfrak{S}\rangle_1, \langle\mathfrak{S}\rangle_2)$ to $\mathbb{Q}$
(3.4) $\mathtt{Index_o} \leftarrow \mathtt{Index_o} + 1$
(4) if $\mathtt{Index_f}$ points to a pair in $\mathbb{Q}$, then go to (2)
(5) stop

Suppose that each of $\mathfrak{C}_1$ and $\mathfrak{C}_2$ has at most n object regions, the length of the $\mathbb{Q}$ would be $C*n$, where C is a constant, the length of the $\mathbb{O}$ would be n^2, the computational complexity of step (1) would be no higher than $O(n)*O(n^2) = O(n^3)$, the computational complexity of step (2) would be no higher than $O(n^5)*C*n = O(n^6)$, the computational complexity of step (3) would be no higher than $(O(n)+O(n)+O(n^2))*O(n^2) = O(n^4)$, the computational complexity of the whole process is $O(n^3)+(O(n^6)+O(n^4))*O(n) = O(n^7)$.

5.7.6 *The Judgement Process*

Recognizing the perceived environment is the judgement of whether the cognitive spectrum of the perceived environment is compatible with the cognitive spectrum of the target environment. The compatibility is determined by the spatial difference between the two cognitive spectrums and the commonsense knowledge of relative stabilities of related object regions.

5.7.6.1 The Spatial Differences

Definition 5.7.18. *Let $\mathbb{O} = [(O_{10}, O_{20}), \ldots, (O_{1n}, O_{2n})]$ be a queue of object region pairs, then $\mathbb{O}$*`.first` *be the set of the first component of each pair in $\mathbb{O}$, $\mathbb{O}$*`.second` *be the set of the second component of each pair in $\mathbb{O}$.*

$$\mathbb{O}.\mathtt{first} = \{O_{1i} | (O_{1i}, O_{2i}) \in \mathbb{O}\}$$
$$\mathbb{O}.\mathtt{second} = \{O_{2j} | (O_{1j}, O_{2j}) \in \mathbb{O}\}$$

Suppose that $\mathbb{O}$ has n elements, then the computational complexity of `.first` and `.second` on $\mathbb{O}$ are no higher than $O(n)$.

For example, let

$$\mathbb{O} = [(ROOM_B, ROOM_C), (BOOKSHELF_B, BOOKSHELF_C), (BOOKS_B, BOOKS_C)]$$
$$\mathbb{O}.\mathtt{first} = \{ROOM_B, BOOKSHELF_B, BOOKS_B\}$$
$$\mathbb{O}.\mathtt{second} = \{ROOM_C, BOOKSHELF_C, BOOKS_C\}$$

Definition 5.7.19. *Let* $\mathfrak{C} =< \langle \mathfrak{S} \rangle, ROOM >$ *be a cognitive spectrum,* $\mathfrak{C}$.basicObjects *be the union of the set* $\{ROOM\}$ *and the set of object regions that are located in* $\mathfrak{S}_{ROOM}(\mathbf{C})$.

$$\mathfrak{C}.\texttt{basicObjects} = \{ROOM\} \cup \mathfrak{S}_{ROOM}(\mathbf{C}).\texttt{Objects}$$

The computational complexity of .basicObjects is a constant.

Definition 5.7.20. *Let* $\mathfrak{C}_1$ *and* $\mathfrak{C}_2$ *be two cognitive spectrums,* $\mathbb{O}$ *be the queue of the mapped object regions. The spatial difference between* $\mathfrak{C}_1$ *and* $\mathfrak{C}_2$*, written as '*$D(\mathfrak{C}_1, \mathfrak{C}_2)$*', is the set of object regions that are not* mapped.

$$D(\mathfrak{C}_1, \mathfrak{C}_2) \stackrel{\text{def}}{=} (\mathfrak{C}_1.\texttt{basicObjects} - \mathbb{O}.\texttt{first}) \cup (\mathfrak{C}_2.\texttt{basicObjects} - \mathbb{O}.\texttt{second})$$

Suppose that each of $\mathfrak{C}_1$ and $\mathfrak{C}_2$ has at most n object regions, then the computational complexity of $D(\mathfrak{C}_1, \mathfrak{C}_2)$ would be no higher than $O(n^7) * O(n) = O(n^8)$.

5.7.6.2 The Degrees of the Compatibility

Definition 5.7.21. *Let* S *be a set of object regions, then* S.rarelyMoved, S.seldomMoved, S.oftenMoved, *and* S.alwaysMoved *are subsets of* S *which collect rarely moved, seldom moved, often moved, and always moved object regions, respectively.*

$$\begin{aligned}
S.\texttt{rarelyMoved} &= \{O | O \in S \wedge O.\texttt{category} = \texttt{rarelyMoved}\} \\
S.\texttt{seldomMoved} &= \{O | O \in S \wedge O.\texttt{category} = \texttt{seldomMoved}\} \\
S.\texttt{oftenMoved} &= \{O | O \in S \wedge O.\texttt{category} = \texttt{oftenMoved}\} \\
S.\texttt{alwaysMoved} &= \{O | O \in S \wedge O.\texttt{category} = \texttt{alwaysMoved}\}
\end{aligned}$$

Suppose that S has n elements, then the computational complexities of .rarelyMoved, .seldomMoved, .oftenMoved, and .alwaysMoved are $O(n)$.

Definition 5.7.22. *Let* $\mathfrak{C}_1$ *and* $\mathfrak{C}_2$ *be two cognitive spectrums,* $D(\mathfrak{C}_1, \mathfrak{C}_2)$ *be the spatial difference between* $\mathfrak{C}_1$ *and* $\mathfrak{C}_2$*. If there are rarely moved objects in* $D(\mathfrak{C}_1, \mathfrak{C}_2)$*, then* $\mathfrak{C}_1$ *and* $\mathfrak{C}_2$ *are* hardly compatible, *written as '*hardlyCompatible($\mathfrak{C}_1, \mathfrak{C}_2$)*'.*

$$\begin{aligned}
H_1 &= \mathfrak{C}_1.\texttt{basicObjects.rarelyMoved} \\
H_2 &= \mathfrak{C}_2.\texttt{basicObjects.rarelyMoved} \\
\texttt{hardlyCompatible}(\mathfrak{C}_1, \mathfrak{C}_2) &\stackrel{\text{def}}{=} (D(\mathfrak{C}_1, \mathfrak{C}_2) \cap (H_1 \cup H_2)) \neq \emptyset
\end{aligned}$$

Suppose that each of $\mathfrak{C}_1$ and $\mathfrak{C}_2$ has at most n object regions, then the computational complexity of hardlyCompatible($\mathfrak{C}_1, \mathfrak{C}_2$) would be no higher than the multiplication of the computational complexity of $D(\mathfrak{C}_1, \mathfrak{C}_2)$ and that of .rarelyMoved. That is, the computational complexity is no higher than $O(n^8) * O(n) = O(n^9)$.

Definition 5.7.23. *Let* $\mathfrak{C}_1$ *and* $\mathfrak{C}_2$ *be two cognitive spectrums,* $D(\mathfrak{C}_1, \mathfrak{C}_2)$ *be the spatial difference between* $\mathfrak{C}_1$ *and* $\mathfrak{C}_2$*. If* $\mathfrak{C}_1$ *and* $\mathfrak{C}_2$ *are not* hardly compatible *and there* *are seldom moved objects in* $D(\mathfrak{C}_1, \mathfrak{C}_2)$*, then* $\mathfrak{C}_1$ *and* $\mathfrak{C}_2$ *are* possibly compatible, *written as* 'possiblyCompatible($\mathfrak{C}_1, \mathfrak{C}_2$)'.

$$P_1 = \mathfrak{C}_1.\texttt{basicObjects.seldomMoved}$$
$$P_2 = \mathfrak{C}_2.\texttt{basicObjects.seldomMoved}$$
$$\texttt{possiblyCompatible}(\mathfrak{C}_1, \mathfrak{C}_2) \stackrel{\text{def}}{=} \neg\texttt{hardlyCompatible}(\mathfrak{C}_1, \mathfrak{C}_2) \wedge (D(\mathfrak{C}_1, \mathfrak{C}_2) \cap (P_1 \cup P_2)) \neq \emptyset$$

Definition 5.7.24. *Let* $\mathfrak{C}_1$ *and* $\mathfrak{C}_2$ *be two cognitive spectrums,* $D(\mathfrak{C}_1, \mathfrak{C}_2)$ *be the spatial difference between* $\mathfrak{C}_1$ *and* $\mathfrak{C}_2$*. If* $\mathfrak{C}_1$ *and* $\mathfrak{C}_2$ *are neither* hardly compatible *nor* possibly compatible *and there are often moved objects in* $D(\mathfrak{C}_1, \mathfrak{C}_2)$*, then* $\mathfrak{C}_1$ *and* $\mathfrak{C}_2$ *are* compatible, *written as* 'Compatible($\mathfrak{C}_1, \mathfrak{C}_2$)'.

$$F_1 = \mathfrak{C}_1.\texttt{basicObjects.oftenMoved}$$
$$F_2 = \mathfrak{C}_2.\texttt{basicObjects.oftenMoved}$$
$$\texttt{Compatible}(\mathfrak{C}_1, \mathfrak{C}_2) \stackrel{\text{def}}{=} \neg\texttt{hardlyCompatible}(\mathfrak{C}_1, \mathfrak{C}_2) \wedge \neg\texttt{possiblyCompatible}(\mathfrak{C}_1, \mathfrak{C}_2) \wedge (D(\mathfrak{C}_1, \mathfrak{C}_2) \cap (F_1 \cup F_2)) \neq \emptyset$$

Definition 5.7.25. *Let* $\mathfrak{C}_1$ *and* $\mathfrak{C}_2$ *be two cognitive spectrums,* $D(\mathfrak{C}_1, \mathfrak{C}_2)$ *be the spatial difference between* $\mathfrak{C}_1$ *and* $\mathfrak{C}_2$*. If* $\mathfrak{C}_1$ *and* $\mathfrak{C}_2$ *are neither* hardly compatible, *nor* possibly compatible, *nor* compatible, *and there are always moved objects in* $D(\mathfrak{C}_1, \mathfrak{C}_2)$*, then* $\mathfrak{C}_1$ *and* $\mathfrak{C}_2$ *are* very compatible, *written as* 'veryCompatible($\mathfrak{C}_1, \mathfrak{C}_2$)'.

$$Q_1 = \mathfrak{C}_1.\texttt{basicObjects.alwaysMoved}$$
$$Q_2 = \mathfrak{C}_2.\texttt{basicObjects.alwaysMoved}$$
$$\texttt{veryCompatible}(\mathfrak{C}_1, \mathfrak{C}_2) \stackrel{\text{def}}{=} \neg\texttt{hardlyCompatible}(\mathfrak{C}_1, \mathfrak{C}_2) \wedge \neg\texttt{possiblyCompatible}(\mathfrak{C}_1, \mathfrak{C}_2) \wedge \neg\texttt{Compatible}(\mathfrak{C}_1, \mathfrak{C}_2) \wedge (D(\mathfrak{C}_1, \mathfrak{C}_2) \cap (Q_1 \cup Q_2)) \neq \emptyset$$

Definition 5.7.26. *Let* $\mathfrak{C}_1$ *and* $\mathfrak{C}_2$ *be two cognitive spectrums,* $D(\mathfrak{C}_1, \mathfrak{C}_2)$ *be the spatial difference between* $\mathfrak{C}_1$ *and* $\mathfrak{C}_2$*. If* $\mathfrak{C}_1$ *and* $\mathfrak{C}_2$ *are neither* hardly compatible, *nor* possibly compatible, *nor* compatible, *nor* very compatible, *then* $\mathfrak{C}_1$ *and* $\mathfrak{C}_2$ *are indeed compatible, written as* 'indeedCompatible($\mathfrak{C}_1, \mathfrak{C}_2$)'.

$$\texttt{indeedCompatible}(\mathfrak{C}_1, \mathfrak{C}_2) \stackrel{\text{def}}{=} \neg\texttt{hardlyCompatible}(\mathfrak{C}_1, \mathfrak{C}_2) \wedge \neg\texttt{possiblyCompatible}(\mathfrak{C}_1, \mathfrak{C}_2) \wedge \neg\texttt{Compatible}(\mathfrak{C}_1, \mathfrak{C}_2) \wedge \neg\texttt{veryCompatible}(\mathfrak{C}_1, \mathfrak{C}_2)$$

Suppose that each of $\mathfrak{C}_1$ and $\mathfrak{C}_2$ has at most n object regions, the computational complexities of possiblyCompatible($\mathfrak{C}_1,\mathfrak{C}_2$), Compatible($\mathfrak{C}_1,\mathfrak{C}_2$), veryCompatible($\mathfrak{C}_1,\mathfrak{C}_2$), and indeedCompatible($\mathfrak{C}_1,\mathfrak{C}_2$) are all no higher than $O(n^9)$.

Theorem 5.7.1. *Let $\mathfrak{C}_1$ and $\mathfrak{C}_2$ be two cognitive spectrums, their five relations be as follows* hardlyCompatible($\mathfrak{C}_1,\mathfrak{C}_2$), possiblyCompatible($\mathfrak{C}_1,\mathfrak{C}_2$), Compatible($\mathfrak{C}_1,\mathfrak{C}_2$), veryCompatible($\mathfrak{C}_1,\mathfrak{C}_2$), *and* indeedCompatible($\mathfrak{C}_1,\mathfrak{C}_2$), *then the five relations are jointly exhaustive and pairwise disjoint.*

$$
\begin{aligned}
&\texttt{hardlyCompatible}(\mathfrak{C}_1,\mathfrak{C}_2) \vee \texttt{possiblyCompatible}(\mathfrak{C}_1,\mathfrak{C}_2) \vee \texttt{Compatible}(\mathfrak{C}_1,\mathfrak{C}_2)\\
&\quad \vee \texttt{veryCompatible}(\mathfrak{C}_1,\mathfrak{C}_2) \vee \texttt{indeedCompatible}(\mathfrak{C}_1,\mathfrak{C}_2) = \texttt{true}\\
&\texttt{hardlyCompatible}(\mathfrak{C}_1,\mathfrak{C}_2) \wedge \texttt{possiblyCompatible}(\mathfrak{C}_1,\mathfrak{C}_2) = \texttt{false}\\
&\texttt{hardlyCompatible}(\mathfrak{C}_1,\mathfrak{C}_2) \wedge \texttt{Compatible}(\mathfrak{C}_1,\mathfrak{C}_2) = \texttt{false}\\
&\texttt{hardlyCompatible}(\mathfrak{C}_1,\mathfrak{C}_2) \wedge \texttt{veryCompatible}(\mathfrak{C}_1,\mathfrak{C}_2) = \texttt{false}\\
&\texttt{hardlyCompatible}(\mathfrak{C}_1,\mathfrak{C}_2) \wedge \texttt{indeedCompatible}(\mathfrak{C}_1,\mathfrak{C}_2) = \texttt{false}\\
&\texttt{possiblyCompatible}(\mathfrak{C}_1,\mathfrak{C}_2) \wedge \texttt{Compatible}(\mathfrak{C}_1,\mathfrak{C}_2) = \texttt{false}\\
&\texttt{possiblyCompatible}(\mathfrak{C}_1,\mathfrak{C}_2) \wedge \texttt{veryCompatible}(\mathfrak{C}_1,\mathfrak{C}_2) = \texttt{false}\\
&\texttt{possiblyCompatible}(\mathfrak{C}_1,\mathfrak{C}_2) \wedge \texttt{indeedCompatible}(\mathfrak{C}_1,\mathfrak{C}_2) = \texttt{false}\\
&\texttt{Compatible}(\mathfrak{C}_1,\mathfrak{C}_2) \wedge \texttt{veryCompatible}(\mathfrak{C}_1,\mathfrak{C}_2) = \texttt{false}\\
&\texttt{Compatible}(\mathfrak{C}_1,\mathfrak{C}_2) \wedge \texttt{indeedCompatible}(\mathfrak{C}_1,\mathfrak{C}_2) = \texttt{false}\\
&\texttt{veryCompatible}(\mathfrak{C}_1,\mathfrak{C}_2) \wedge \texttt{indeedCompatible}(\mathfrak{C}_1,\mathfrak{C}_2) = \texttt{false}
\end{aligned}
$$

5.7.6.3 Making a Judgement in Everyday Life

Definition 5.7.27. *Let $\mathfrak{C}_1$ and $\mathfrak{C}_2$ be two cognitive spectrums, $\mathfrak{C}_1$ is rarely recognized as $\mathfrak{C}_2$, written as* 'hardlyIs($\mathfrak{C}_1,\mathfrak{C}_2$)', *if* hardlyCompatible($\mathfrak{C}_1,\mathfrak{C}_2$).

$$\texttt{hardlyIs}(\mathfrak{C}_1,\mathfrak{C}_2) \stackrel{\text{def}}{=} \texttt{hardlyCompatible}(\mathfrak{C}_1,\mathfrak{C}_2)$$

Definition 5.7.28. *Let $\mathfrak{C}_1$ and $\mathfrak{C}_2$ be two cognitive spectrums, $\mathfrak{C}_1$ might be recognized as $\mathfrak{C}_2$, written as* 'mightBe($\mathfrak{C}_1,\mathfrak{C}_2$)', *if* possiblyCompatible($\mathfrak{C}_1,\mathfrak{C}_2$).

$$\texttt{mightBe}(\mathfrak{C}_1,\mathfrak{C}_2) \stackrel{\text{def}}{=} \texttt{possiblyCompatible}(\mathfrak{C}_1,\mathfrak{C}_2)$$

Definition 5.7.29. *Let $\mathfrak{C}_1$ and $\mathfrak{C}_2$ be two cognitive spectrums, $\mathfrak{C}_1$ is recognized as $\mathfrak{C}_2$, written as* 'Is($\mathfrak{C}_1,\mathfrak{C}_2$)', *if* Compatible($\mathfrak{C}_1,\mathfrak{C}_2$).

$$\texttt{Is}(\mathfrak{C}_1,\mathfrak{C}_2) \stackrel{\text{def}}{=} \texttt{Compatible}(\mathfrak{C}_1,\mathfrak{C}_2)$$

Definition 5.7.30. *Let $\mathfrak{C}_1$ and $\mathfrak{C}_2$ be two cognitive spectrums, $\mathfrak{C}_1$ is exactly $\mathfrak{C}_2$, written as* 'exactlyIs($\mathfrak{C}_1,\mathfrak{C}_2$)', *if* veryCompatible($\mathfrak{C}_1,\mathfrak{C}_2$).

$$\texttt{exactlyIs}(\mathfrak{C}_1,\mathfrak{C}_2) \stackrel{\text{def}}{=} \texttt{veryCompatible}(\mathfrak{C}_1,\mathfrak{C}_2)$$

Definition 5.7.31. *Let* $\mathfrak{C}_1$ *and* $\mathfrak{C}_2$ *be two cognitive spectrums,* $\mathfrak{C}_1$ *is indeed* $\mathfrak{C}_2$*, written as* '$\mathtt{indeedIs}(\mathfrak{C}_1,\mathfrak{C}_2)$'*, if* $\mathtt{indeedCompatible}(\mathfrak{C}_1,\mathfrak{C}_2)$.

$$\mathtt{indeedIs}(\mathfrak{C}_1,\mathfrak{C}_2) \stackrel{\text{def}}{=} \mathtt{indeedCompatible}(\mathfrak{C}_1,\mathfrak{C}_2)$$

5.8 The Mereotopological Formalism of the Theory of Cognitive Prism

This chapter formalized The Theory of Cognitive Prism in the notions of *regions* (including *object regions*, *side regions*, and *constructed regions*), and *fiat containers*. The *fiat* container is constructed through the connectedness relation. *Locations* of object regions are defined in the notion of *fiat* containers based on the principle of selecting *fiat* containers. The cognitive spectrum of an indoor spatial environment is formalized by the pair of a set of *fiat* containers and an object region representing the room of the indoor environment. Relations between two cognitive spectrums are addressed based on the *categorically the same* relation between *object regions*. Two rooms in two cognitive spectrums are *mapped*, if they are *categorically the same*; objects inside of room are *mapped*, if they are *categorically the same* and their locations are *mapped*. The compatibility between two cognitive spectrums are determined by the un-mapped object regions and their relative stabilities. Recognizing variable spatial environments in the everyday-life situation is interpreted as the compatibility between the cognitive spectrum of the perceived environment and that of the target environment.

Chapter 6
A List Representation of Recognizing Indoor Vista Spatial Environments: The LIVE Model

You know my methods. Apply them!
— Sherlock Holmes

This chapter introduces a symbolic representation system, the LIVE model. Section 6.1 briefly introduces the general structure of the LIVE model; section 6.2 introduces the usage of the LIVE model; section 6.3-6.6 presents the list representation of indoor vista spatial environment that pertains to the formalism in Chapter 5; section 6.7 examines of the relationship between the principle of reference in spatial linguistic description and the principle of selecting cognitive reference objects; section 6.8 presents the representation of the recognition process; section 6.9 presents the simulation result of Mr. Bertel's apartment scenario of Chapter 1 in the LIVE model.

6.1 The General Architecture of the LIVE Model

The general architecture of the LIVE model is shown in Figure 6.1. It has five submodels: The furniture system, the drawing system, the configuration files, the view system, and the comparison system.

The furniture system stores all the furniture information that the LIVE model represents. The configuration files store all the symbolic representations of configurations. The view system provides a graphical interface for the symbolic representation of configurations. The drawing system provides a graphical interface to create or modify configurations. The comparison system judges the compatibility of two selected configurations.

The furniture system is the basic system. It provides furniture class information for the view system, the drawing system, and the comparison system. The view system provides an easy way for the draw system to create new configurations by modifying existing one.

T. Dong: Recognizing Variable Environments, SCI 388, pp. 79–95.
springerlink.com

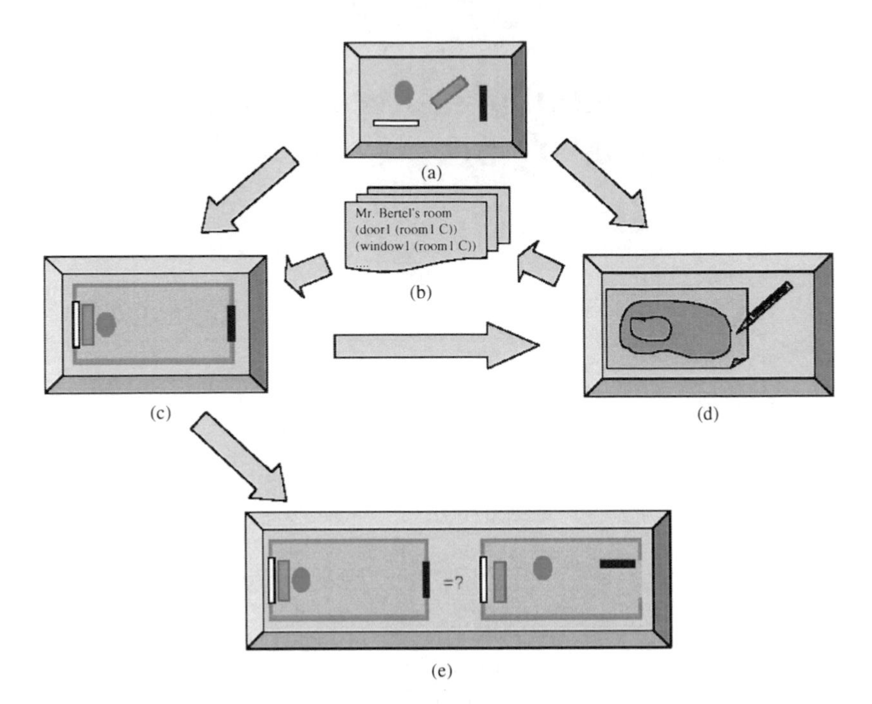

Fig. 6.1 The general architecture of the LIVE model: (a) The furniture system, (b) configuration files, (c) the view system, (d) the drawing system, (e) the comparison system. Arrows represent information flow

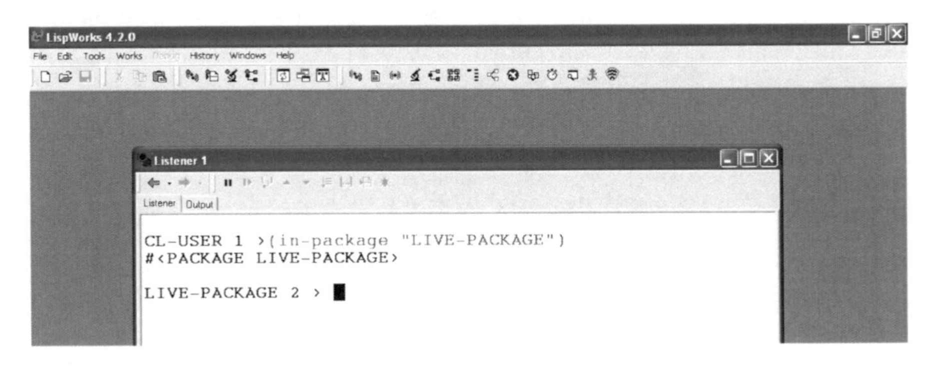

Fig. 6.2 Before starting the LIVE model, you had better go into the "`LIVE-PACKAGE`"

6.2 How to Start the LIVE Model?

The LIVE model has been implemented in LispWorks4.2 both on the Linux Susie 6.3 platform and on the Windows XP professional platform.

To start the LIVE model, the "'`loader.lisp`" should be loaded firstly; then, type "`(in-package "LIVE-PACKAGE")`", shown in Figure 6.2; thirdly, type "`(load-live)`" to load all the "`.fsl`" files of the LIVE model; at last, type

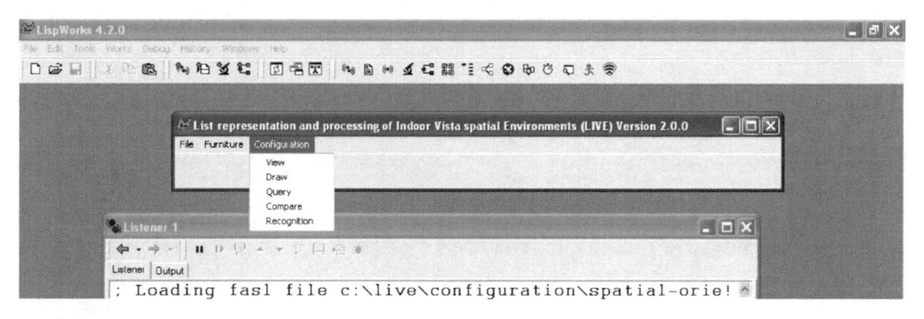

Fig. 6.3 The main menu of the LIVE model

"`(start-live-demo)`" and you will see the main menu of the LIVE model, shown in Figure 6.3.

6.3 The Furniture System

If you click onto "`furniture`" on the main menu and then click on "`view`", there will be a furniture view window prompted. This window displays all the furniture (e.g., doors, windows, couches, etc.) in the LIVE model both symbolically and diagrammatically, as shown in Figure 6.4.

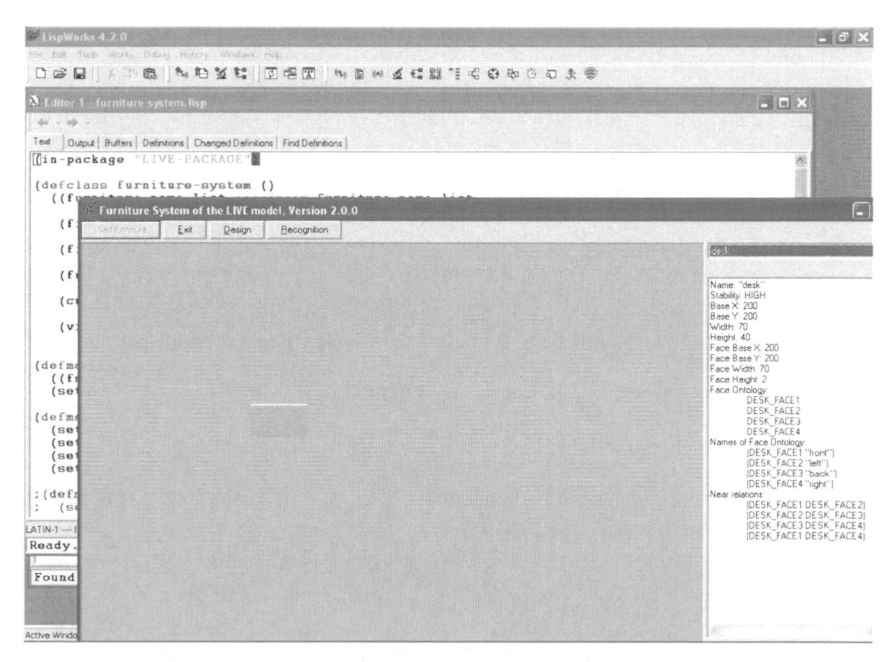

Fig. 6.4 The view window of the LIVE furniture system

In the LIVE model, a piece of furniture is represented by an instance of a class. The class has the category knowledge about the object, such as the name of this category, the default values of the degree of the relative stability, sides, etc.

6.3.0.1 Class Names

In the LIVE model, class names are constructed by adding `c_` before the name in natural language. For example, the object class name of the room is `c_room`. The LIVE model has 15 classes representing 15 preferable categories: `ROOM` (the `c_room` class), `WINDOW` (the `c_window` class), `DOOR` (the `c_door` class), `WRITING– DESK` (the `c_desk` class), `BOOKSHELF` (the `c_shelf` class), `COUCH` (the `c_couch` class), `SMALLCOUCH` (the `c_smallcouch` class), `BALLOON` (the `c_balloon` class), `TEATABLE` (the `c_table` class), `CHAIR` (the `c_chair` class), `TABLE` (the `c_table` class), `BOOK` (the `c_book` class), `CUP` (the `c_cup` class), `PICTURE` (the `c_picture` class), `FLOWER` (the `c_flower` class).

For the convenience of creating new environments, the `c_room` class in the LIVE model does not represent the `ROOM` category in Chapter 4. The `c_room` class does not include doors and windows, so that a new empty room environment can be easily created by putting doors and windows into difference sides of a `c_room` instance.

6.3.0.2 Object Names

In the LIVE model, an object name has two parts: The first part is the name in English, the second part is a number which is used to distinguish different instances of the same class in one configuration. For example, names of windows can be `window1`, `window2`.

6.3.0.3 The Relative Stability

The LIVE model distinguishes four levels of stability: `highest`, `high`, `low`, and `lowest`. The `highest` stability represents the `rarelyMoved` stability; the `high` stability represents the `seldomMoved` stability; the `low` stability represents the `oftenMoved` stability; and the `lowest` stability represents the `alwayslyMoved` stability.

6.3.0.4 Face Information

As the LIVE model does not represent observers, it does not represent the *fiat* projection mechanism. It distinguishes two kinds of objects: Objects that have an intrinsic reference framework and objects that do not have an intrinsic reference framework. Objects with an intrinsic reference framework have four faces: `<category name>_face1`, `<category name>_face2`, `<category name> _face3`, and `<category name>_face4`. Objects that do not have an intrinsic reference framework only have one face:

`<category name>_face1`. For example, main slots of `c_window` are as follows:

```
(defclass c_window (c_furniture_rectangle)
 % this class is inherited from
 % the c_furniture_rectangle class
   (name :accessor name
         % the default name of a window
        :initform "window1")
   (stability :accessor stability
         % a window is rarely moved
             :initform 'highest)
   (faces :accessor faces
         % windows have intrinsic orientation
         % reference frameworks
         :initform '(window_face1 window_face2
                     window_face3 window_face4))
   (face-names :accessor face-names
         % the linguistic descriptions of
         % each face of a window
              :initform '((window_face1 "front")
                         (window_face2 "left")
                         (window_face3 "outside")
                         (window_face4 "right"))))
```

In the LIVE model, all objects are stored in a hash-table structure which is indexed by the name of the object.

6.4 Configurations in the LIVE Model

In LIVE model a *configuration* refers to a set of list representation of the spatial linguistic description of an indoor spatial environment.

6.4.1 Spatial Relations in the LIVE Model

The `Object-Object relation` and the `Object-Face relation` in the LIVE model represent distance (including connectedness) relations and orientation relations, respectively.

6.4.1.1 The Object-Object Relation

The connectedness or distance *fiat* container is symbolically represented by a two-element list. The first element is the name of the anchor object of the *fiat* container, the second element is a symbol representing the connectedness relation or a distance relation. Let `C`[1] be the symbol for the connectedness relation, `NR` be the symbol for the distance relation of *near*, `FR` be the symbol for the distance relation of *far*, then `(bookshelf1 NR)` represents the *near fiat* container with `bookshelf1` as the anchor object.

The distance location(including the connectedness relation) of the object is symbolically represented by an `Object-Object-Relation` list. It has two elements. The first element is the name of the object. The second element is the list whose element is the list representing a connectedness or distance *fiat* container. For example, `(desk1 ((room1 C) (window1 NR) (shelf1 NR)))` is the `Object-Object-Relation` list of the object `desk1`. It represents that `desk1` is located in the *fiat* containers of `(room1 C)`, `(window1 NR)`, and `(shelf1 NR)`.

6.4.1.2 The Object-Face Relation

The orientation *fiat* container is symbolically represented by a three-element list. The first element is the name of the anchor object of the *fiat* container, the second element is the side of the anchor object such that any object located in the *fiat* container is nearer to this side, the third element `NRR` represents the nearer relation. For example, the front *fiat* container of the bookshelf `bookshelf1` is symbolically represented by `(bookshelf1 BOOKSHELF_FACE1 NRR)`.

An object's orientation location is represented by the `Object-Face-Relation` list. It has two elements. The first element is the name of an object. The second element is a list. Each element of this list is also a list that represents an orientation *fiat* container in which the object is located. For example, the `Object-Face-Relation` list `(balloon1 ((bookshelf1 BOOKSHELF_FACE1 NRR)))` represents `balloon1`'s orientation. The second element represents that `balloon1` is located nearer to the side `BOOKSHELF_FACE1` of `bookshelf1` than to its other sides.

6.4.2 The Drawing System

The drawing system provides graphical interfaces to create new configurations or modify already existing configurations. The drawing system has three sub-systems: The furniture draw panel sub-system, the furniture rotation sub-system, and the spatial relation sub-system. The furniture draw panel is an interface for adding or removing furniture; the furniture rotation panel is an interface for rotating furniture 90°, 180°, or 270°, the qualitative spatial panel is an interface for setting qualitative spatial relations between pieces of furniture, as shown in Figure 6.5.

[1] In the current version of the LIVE model, the connectedness relation `C` is specified into seven of the RCC-8 relations, namely, `EC`, `PO`, `TPP`, `NTPP`, `EQ`, $\texttt{TPP}^{-1}$, and $\texttt{NTPP}^{-1}$.

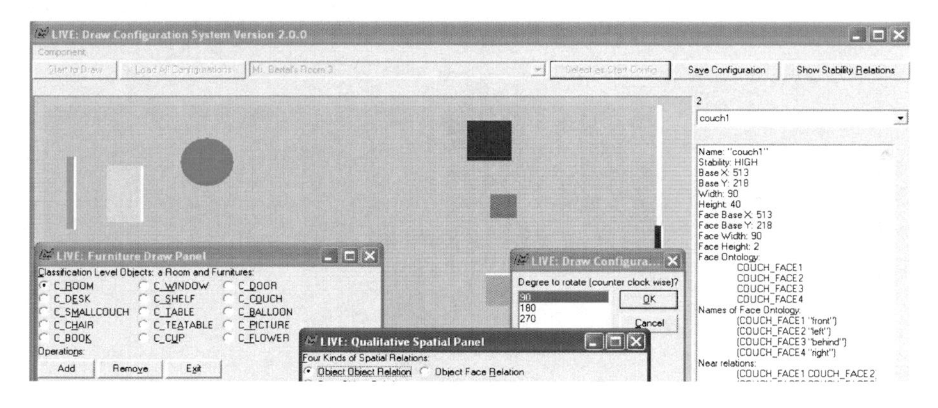

Fig. 6.5 The drawing system creates a new configuration

6.5 The Configuration File

Symbolic representations of configurations are stored in a file, which includes symbolic representations of furniture (shape, color, sides, etc.) and their spatial relations. For example, the list representation of Mr. Bertel's apartment with no furniture is shown as follows.

```
(OBJECT-OBJECT-RELATION (|window1|
                           ((|room1| C) (|door1| FR)))
                          (|door1|
                           ((|room1| C) (|window1| FR))))
(OBJECT-FACE-RELATION (|window1|
                           ((|room1| ROOM_FACE3 NRR)))
                        (|door1|
                           ((|room1| ROOM_FACE1 NRR))))
```

6.6 The View System

The view system displays a configuration file diagrammatically. For example, Mr. Bertel's decorated apartment is shown in Figure 6.6.

6.7 Testing the Principle of Selecting *fiat* Containers

The principle of selecting *fiat* containers is tested in the LIVE model by checking the diagram of the reference relation between the location object and the reference object. The method is as follows: If an object `obj1` is referenced to an object `obj2` and `obj2` is not reference to `obj1`, then there will be a line drawn between `obj1` and `obj2` and `obj2` is located a bit higher than `obj1`. The line represents the reference relation between two objects; the location in height represents the relative stability among objects — the higher an object is located, the more stable it is. If an

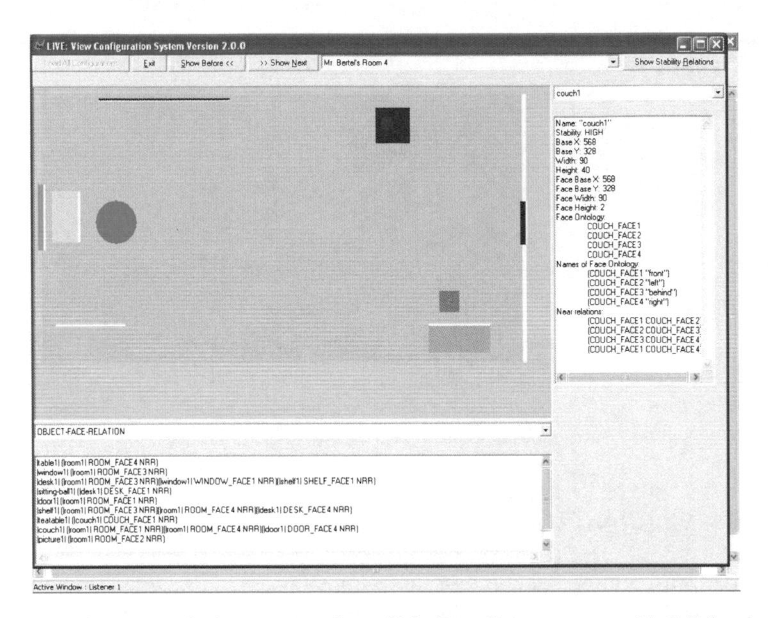

Fig. 6.6 The diagrammatical representation of Mr. Bertel's apartment with full furniture and decoration

object `obj1` is referenced to an object `obj2`, and `obj2` is also reference to `obj1`, then `obj2` is located at the same height as `obj1`, and there is no line between them.

The diagram including all the reference relations of a configuration is a partial hierarchical structure, called "the linguistic reference hierarchy". The view system displays the linguistic reference hierarchy diagrammatically. Click on "`Show Sta- bility Relation`" button in the window of Figure 6.7, and a window to view three hierarchical structures relating to this configuration will be prompted, as shown in Figure 6.7. For example, click the button of "`Show Linguistic Relation`", and the linguistic references of the configuration in Figure 6.7 will be shown in Figure 6.8.

A linguistic reference hierarchy may have triangles. For example, an object `obj1` is referenced to the second object `obj2`, a third object `obj3` is referenced to `obj2`, `obj1` is referenced to `obj3`, and `obj3` is located lower than `obj2` and higher than `obj1`, then there will be a triangle among `obj1`, `obj2` and `obj3`. If an object `obj1` is referenced to `obj2` and there is no third object `obj3` locating lower than `obj2` and higher than `obj1` (This guarantees that `obj3` is not referenced to `obj1` and that `obj2` is not referenced to `obj3`.) such that `obj1` is referenced to `obj3` and `obj3` is referenced to `obj2`, then the reference between `obj1` and `obj2` is called "a direct reference", or else it is called "an indirect reference". Removing all the indirect reference relations from a linguistic reference hierarchy results a diagram called "the hierarchical structure of reference objects". A click on the button of "`Show Lattice Relation`" will prompt the hierarchical structure of direct reference relations of the configuration in Figure 6.6, shown in Figure 6.9.

Fig. 6.7 The window framework of three hierarchical structures

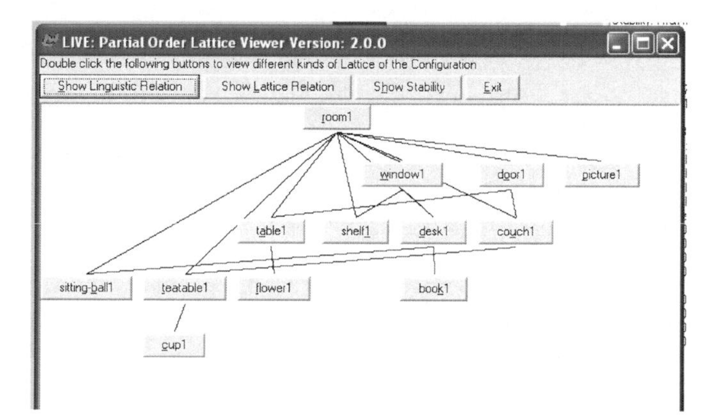

Fig. 6.8 The diagrammatic representation of linguistic reference relations of objects in Mr. Bertel's apartment

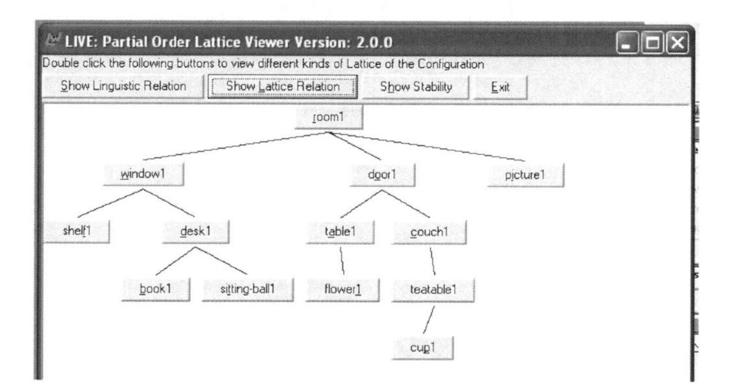

Fig. 6.9 The diagrammatic representation of partial hierarchical structure of reference objects of Mr. Bertel's apartment

The result of the diagram in this example is consistent with the principle of selecting *fiat* containers as follows: The direct reference object is nearer to the location object than indirect reference objects; always moved objects (except `picture1`) are located lower in the diagram than the often moved objects; often moved objects are located lower than the seldom moved objects; seldom moved objects are located lower than the rarely moved objects. The result of the diagram in this example is inconsistent with the principle of selecting relative spaces as follows: Although `room1`, `window1` and `door1` are rarely moved objects, `room1` is located higher than `window1` and `door1`; although `picture1` is an always moved object[2], it is located at the same height as `window1` and `door1` which are rarely moved objects.

The reason for the inconsistency among `room1`, `window1`, and `door1` is that although `room1`, `window1` and `door1` are all rarely moved objects, the location change of `room1` is even much harder than that of `window1` and `door1`. That is, the pre-assumed degree of "rarely moved" granularity is a bit too coarse for `room1`, `window1` and `door1`. To meet the pre-assumed degree of "rarely moved" granularity, the partial hierarchical structure of reference objects is modified as follows: The rarely moved objects of a configuration are set at the same height. The original root is replaced by the name of the configuration. Click the button of "`Show stability`" to prompt the hierarchical structure of stability of the configuration in Figure 6.6 (shown in Figure 6.10). The modified structure is called "the partial order lattice of stabilities". The inconsistency of `picture1` lies in the fact that it is hung on the wall, which promotes its relative stability. In the partial order lattice of stabilities, the picture hung on the wall is of the same relative stability as the shelf, the desk, the table, and the couch.

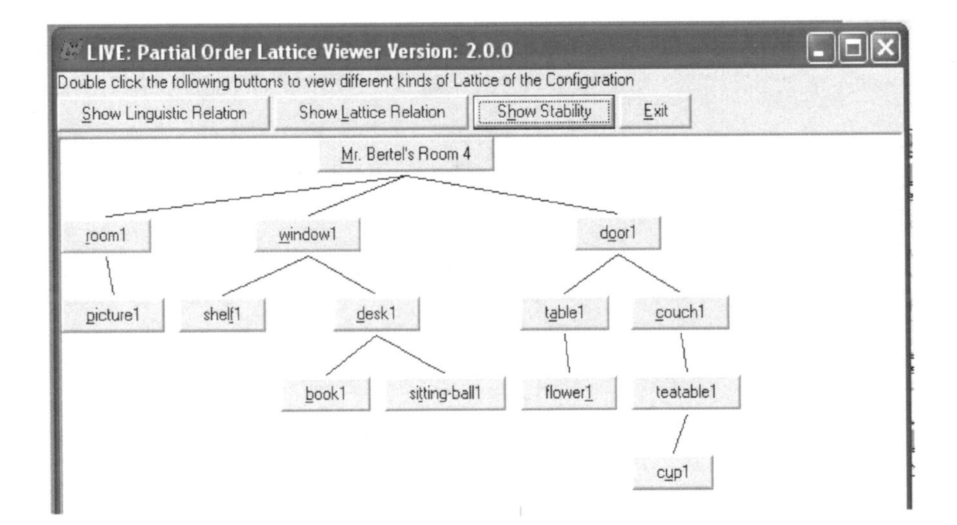

Fig. 6.10 The diagrammatic representation of partial order lattice of relative stability of Mr. Bertel's apartment

[2] As in the LIVE model, it is not assumed that pictures must be hung on the wall, therefore, the relative stability of a picture is set to always moved.

6.8 The Comparison System

The comparison system compares the two selected configurations. The first selected configuration represents the configuration in the mind; the second selected configuration represents the current perceived environment. The comparison system makes a judgment on to which degree the second configuration is recognized as the first one. Figure 6.11 shows the window frame of the comparison configuration system in the LIVE model. The functions of the five buttons are as follows: `Select the first configuration` and `Select the second configuration` buttons are used to select two configurations. After clicking each of them, the view configuration system will generate a configuration view interface for the user to select a configuration. Clicking the `Compare two configurations` button will start the recognition process. It first finds all the spatial differences between the two configurations. Then, it gives a judgment on whether it is the target one based on spatial differences and the relative stabilities of objects.

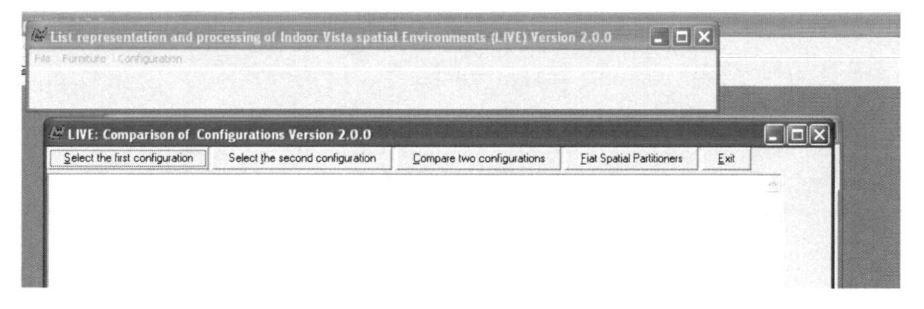

Fig. 6.11 The comparison system in the LIVE model

6.8.1 The Main Structure

The comparison system is the key system in the LIVE model. The three main classes of this sub-system are `compare-configuration-sys` class, `select-con-+fig-viewer`, and `compare-viewer` class.

```
(defclass compare-configuration-sys ()
  ((config-1 :accessor config-1
             :initform nil)
   (lattice-1 :accessor lattice-1
              :initform nil)
   (config-view-1 :accessor config-view-1
                  :initform nil)
   (config-2 :accessor config-2
             :initform nil)
   (lattice-2 :accessor lattice-2
             :initform nil)
   (config-view-2 :accessor config-view-2
                  :initform nil)
```

```
      (fiat-spatial-partition
                    :accessor fiat-spatial-partition
                    :initform nil)
      (fiat-spatial-partition-viewer
             :accessor fiat-spatial-partition-viewer
             :initform nil)
      (result :accessor result
             :initform nil)
      (log-str :accessor log-str
              :initform nil)))
```

The `compare-configuration-sys` class stores all the information that is needed for the comparison of two configurations. The `config-1` and the `config-2` slots store the first and the second configurations, respectively; the `lattice-1` and the `lattice-2` store the partial order lattice of stability in the first configuration and the second configuration respectively. They are constructed based on spatial relations of the two configurations and represent the hierarchy of the cognitive spectrum in Chapter 5. The `config-view-1` and the `config-view-2` slots are the graphical views of the first and the second configurations, respectively. The `fiat-spatial-partition` slot stores the mapped *fiat* containers hierarchically. This structure is created in the recognition process. The `fiat-spatial- +partition-view` slot refers to the graphical view of the mapped *fiat* containers. The `result` stores the result of the comparison process. The `log-str` stores structured linguistic expressions about the comparison process.

6.8.2 The Structure of Mapped **fiat** *Containers*

In the LIVE model, the mapped *fiat* containers are hierarchically constructed in the recognition process. An instance of `fiat-spatial-partition` class represents a mapped *fiat* container. The slots of `fiat-spatial-partition` class are as follows.

```
   (defclass fiat-spatial-partition ()
     ((root-node-rem :accessor root-node-rem
                     :initform nil)
      (root-node-per :accessor root-node-per
                     :initform nil)
      (loc-partition :accessor loc-partition
                     :initform nil)
      (children :accessor children
                :initform nil)))
```

The `root-node-rem` slot and the `root-node-per` slot store objects in mind and objects perceived such that they are located in mapped *fiat* containers. If two rooms are categorically the same, then they are mapped and `root-node-rem` holds the object list `(room1)` of the remembered configuration, `root-node-per` holds the object list `(room1)` of the perceived configuration. The `loc-partition` slot is a hash-table of the distance (including connectedness) *fiat* container that uses distance relations as indexes. The mapped object and the distance relation represent a mapped distance *fiat* container which is an instance of the class `LocationObject`, whose slots of this class are shown as follows.

```
(defclass LocationObject ()
  ((all-rem :accessor all-rem
            :initform nil)
   (all-per :accessor all-per
            :initform nil)
   (orien-partition :accessor orien-partition
                    :initform nil)
                    ;hashtable faces-->PostureObject ))
```

The `all-rem` and the `all-per` slots in `LocationObject` holds objects in the remembered configuration and objects in perceived configuration that are located in mapped distance (including connectedness) *fiat* containers, respectively. Their locations are specified by the orientation *fiat* containers. The `orien-partition` slot holds a hash table of mapped orientation *fiat* containers. This hash table uses side names of the mapped object in `root-node-rem` and `root-node-per` as indexes and creates an instance of the `PostureObject`[3] class.

All object list pairs `all-rem` and `all-per` are checked to see whether there are mapped objects. Instances of `fiat-spatial-partition` class will be created for mapped objects in the earlier created instances of `fiat-spatial-parti- +tion` class. The newly created instance will be appended to the `children` slot of the mother instance. Figure 6.12 shows the instance of `fiat-spatial-partition` of recognizing Mr. Bertel's apartment. The connectedness relation `C` is specified into `PO`, `TPP`, `PP` representing *partially overlapped*, *tangential proper part*, *proper part* relations. `ORO.faceX` represents an orientation *fiat* container such that objects in it are nearer to `faceX` of the reference object than to its other sides. `GES.faceX` represents a posture relation[4].

6.9 The Simulation of Recognizing Mr. Bertel's Apartment

This section presents the simulation results of the Mr. Bertel's new home scenario in Chapter 1.

When Mr. Bertel's mother came to see her son's new home for the first time, she saw the apartment with full furniture and decoration. In LIVE model this configuration is shown in Figure 6.13.

On the next day after Mr. Certel's visiting, Mr. Bertel's mother came to her son's apartment again. She noticed the difference of the location of the balloon, shown in

[3] The posture relation is explained in (Dong, 2005, p.354). Discussing the `PostureObject` class is beyond the scope here.

[4] see note 3, on page 91.

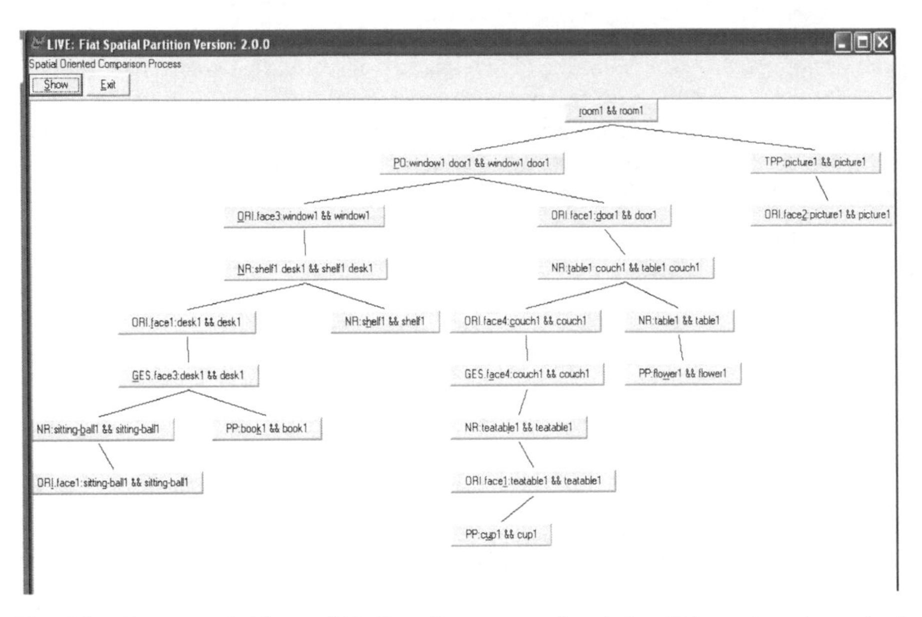

Fig. 6.12 The mapped objects of Mr. Bertel's apartment in mind and his apartment perceived are listed in the mapped *fiat* containers. Objects left and right of "&&" are *mapped*

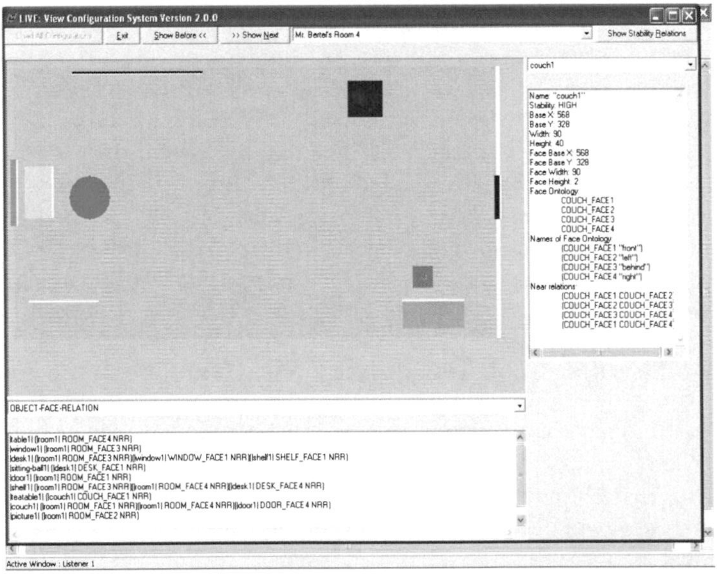

Fig. 6.13 The view system shows the diagrammatic representation of Mr. Bertel's apartment with full furniture and decoration

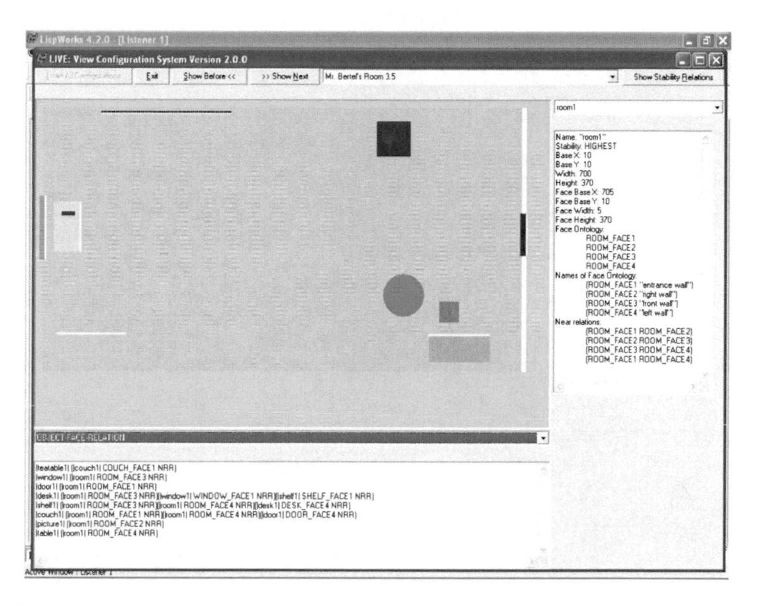

Fig. 6.14 The view system shows the diagrammatic representation of Mr. Bertel's apartment after Mr. Certel's first visiting who moved the balloon to the tea-table

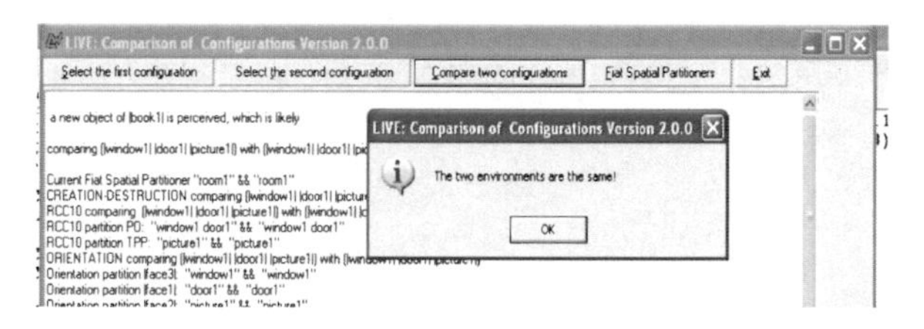

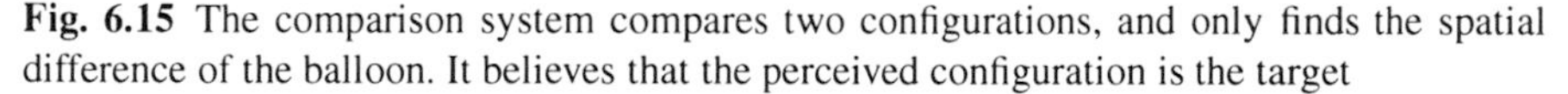

Fig. 6.15 The comparison system compares two configurations, and only finds the spatial difference of the balloon. It believes that the perceived configuration is the target

Figure 6.14. However, she still recognized it her son's apartment, she just mumbled: Why did her son put the balloon to the tea-table? The comparison system of the LIVE model compares the above two configurations and reports "The two environments are the same!", shown in Figure 6.15.

On the eighth day the flowers are withered and thrown away. All books are put on the bookshelf. In the evening Mr. Bertel moves his writing-desk to the right side of the bookshelf to make room for a party. The balloon is put near the tea-table, which is now at the right hand side of the door at the corner. In the LIVE model, Mr. Bertel's apartment after the party is shown in Figure 6.16. On the ninth day Mr. Bertel's mother came again. This time she went into Mr. Certel's apartment by mistake. In LIVE model the configuration of Mr. Certel's apartment is shown in Figure 6.17. Although Mr. Certel's apartment was very similar to Mr. Bertel's apartment, Mr. Bertel's mother noticed that she was not in her son's home and left the

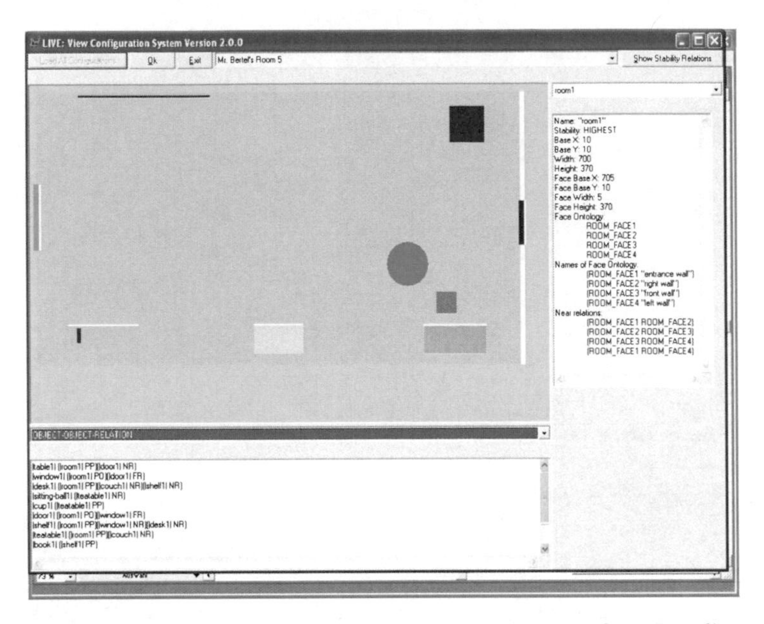

Fig. 6.16 The view system shows the diagrammatic representation of Mr. Bertel's apartment with the after-party layout

Fig. 6.17 The view system shows the diagrammatic representation of Mr. Certel's apartment

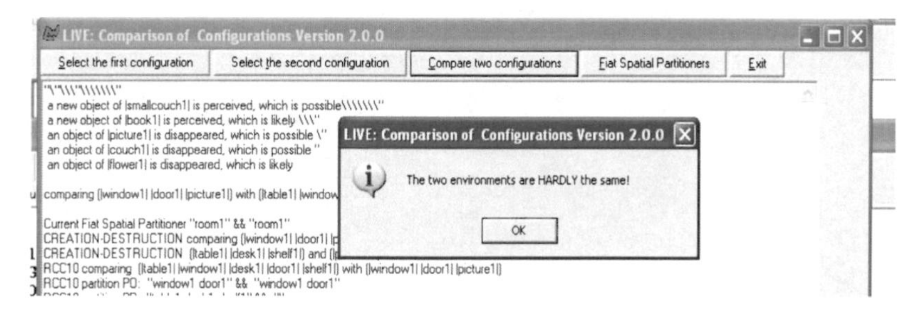

Fig. 6.18 The comparison system of the LIVE model judges that Mr. Certel's apartment and Mr. Bertel's apartment are HARDLY the same

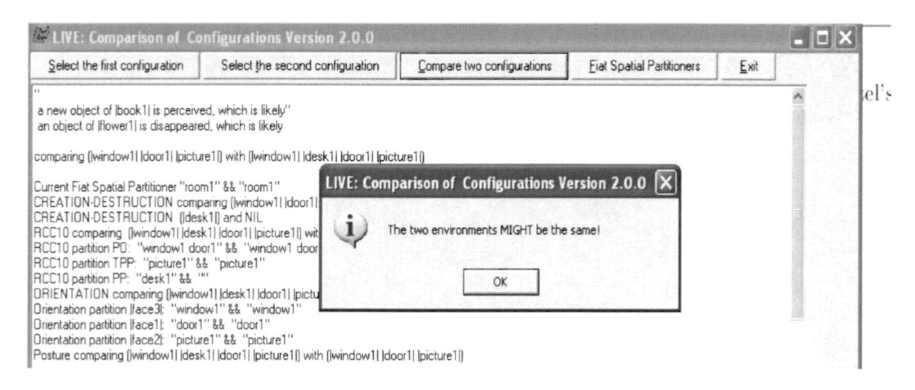

Fig. 6.19 The comparison system of the LIVE model judges that Mr. Bertel's apartment with a after-party layout might be Mr. Bertel's apartment

apartment. The comparison system of the LIVE model compares the two configurations and reports "The two environments are HARDLY the same!", shown in Figure 6.18. When she went to Mr. Bertel's apartment, She found that the writing-desk was located between the bookshelf and the couch and that the table and the balloon were located differently as she expected, she wondered for a while and accepted it as her son's apartment. The comparison system of the LIVE model results "The two environments MIGHT BE the same!", shown in Figure 6.19.

Chapter 7
Conclusions, Evaluations, Discussions, and Future Work

"道可道，非常道；名可名，非常名。"
老子 •《道德经》

"The way that can be told of is not an unvarying way;
The names that can be named are not unvarying names"
Lao Tzu << Tao Te Ching>>

7.1 Conclusions

This work presented a computational theory of recognizing spatial environments — The Theory of Cognitive Prism. The theory assumed that people are able to categorize spatial objects through perception. It proposed the commonsense understanding of distance relations and orientation relations between extended objects and the commonsense knowledge of relative stabilities of extended objects. The knowledge of relative stability is used in the reference ordering between two extended objects following the principle of selecting cognitive reference objects. A snapshot view in the mind, "cognitive spectrum", is represented by nested relative spaces such that relative spaces of less stable objects are nested in relative spaces of more stable objects. The compatibility relationship between two cognitive spectrums is interpreted as the ease of the transformation from one cognitive spectrum to the other. The compatibility is determined by the relative stabilities of the un-mapped objects between two cognitive spectrums. The task of recognizing spatial environments in everyday life is interpreted as the judgement on the compatibility between the perceived environment and the target environment. The theory is mereotopologically formalized. The LIVE model symbolically implemented the main part of The Theory of Cognitive Prism. It tested the principle of selecting *fiat* containers, and symbolically simulated the Mr. Bertel's apartment scenario. The Theory of Cognitive Prism

T. Dong: Recognizing Variable Environments, SCI 388, pp. 97–102.
springerlink.com

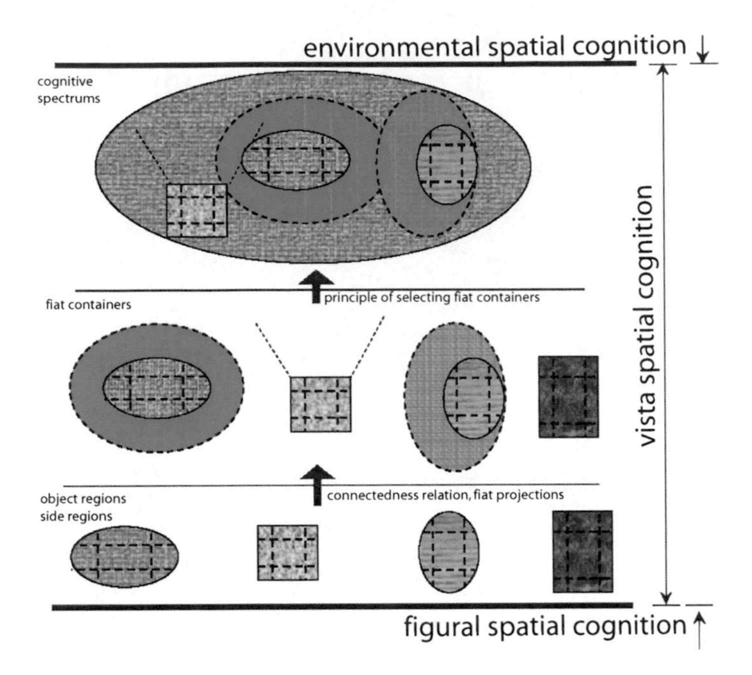

Fig. 7.1 The framework of vista spatial cognition

outlines the research field of *vista spatial cognition* which bridges the research field of figural spatial cognition and environmental spatial cognition, shown in Figure 7.1. Vista spatial cognition starts from the result of figural spatial cognition – knowledge of extended objects, researches relations (e.g., spatial relations, temporal relations, reference relations, etc.) among extended objects, and results in properties (e.g., existence, change, etc.) of vista spatial environments which are primitive for the environmental spatial cognition.

7.2 Evaluation on the Representation and the Reasoning

7.2.1 Obtainable by Cognitive Agents

The first criterion of a spatial representation that can be used by a cognitive agent is that the representation shall be obtainable by the sensors of the cognitive agent. It is assumed in the Theory of Cognitive Prism that a cognitive agent is able to recognize extended objects, make spatial extensions, and have the commonsense knowledge of relative stabilities of objects. With these three conditions, it is described in detail in Chapter 4 how the representation of a spatial environment, called "the cognitive spectrum", is constructed.

7.2.2 Meaningful to Languages

Cognitive agents need to make communications. The second criterion of a spatial representation is that the representation shall provide a systematical way to give meanings of the spatial linguistic descriptions, so that cognitive agents can exchange spatial information through language.

A piece of spatial linguistic description is a linear order of a location object, a reference object, and a spatial relation such that objects of less relative stability appear first. For example, *that the cup is on the table* is a linear order of *the cup*, *the table*, and the relation *on*; *the cup* is less stable than *the table*, therefore, in linguistic description *the cup* appears before *the table* as follows: *the cup is on the table*. In the Theory of Cognitive Prism an object which is located in a relative space has the meaning as follows: This object is relatively less stable than the anchor object of this relative space; this object is relative to the anchor object by this relative space.

In the Theory of Cognitive Prism three kinds of spatial relations (topological, distance, and orientation) are explained based on the connectedness relation. If there are only these three kinds of spatial relations for people (as claimed by Piaget and Inhelder (1948)), it would be a piece of future work for linguists to find out explanations for spatial relations in natural languages that are based on the connectedness relation.

7.2.3 Representing Distortions

Cognitive agents acquire spatial knowledge by interacting with spatial environments using their sensors. Their knowledge of spatial environment inevitably has distortions. The third criterion of a spatial representation is that the representation shall be able to represent spatial distortions. In the Theory of Cognitive Prism objects are categorically recognized. That is, objects of the same category are equivalent and indistinguishable in isolation. Therefore, objects of the same category are believed to be of the same size (the mereotological formalism of object size is presented in Appendix A).

The distortion in spatial distance can result from subjectively putting objects of different sizes into the same category. For example, Sadalla and Magel (1980) reported that the qualitative distance of a route is positively related to the number of right-angle turns distributed along the route and that a route with more turns is estimated as longer than a route of equivalent length with fewer turns. In this case route segments between two consecutive right-angle turns are believed of the same category, and therefore route segments are believed of the same length. Consequently, the subjective distance of a route is positively related to the number of such route segments. A systematical research on the relationship between object categorizations and spatial distortions would be a future work for cognitive psychologists.

7.2.4 Computable

A computational theory shall be computable, so that it can be used by cognitive agents efficiently. It is roughly estimated in Chapter 5 that the computational complexity of the process of recognizing spatial environments in the Theory of Cognitive Prism is 'P' (polynomial to the input size).

7.3 Discussions and Future Work

7.3.1 The Granularity of Object Categorizations

One of the starting points of The Theory of Cognitive Prism is that objects have been categorized. Given the scene of a spatial environment, the cognitive agent has recognized objects in the scene and categorized them at a certain granularity, such as a writing-desk, a table, a couch, etc. The granularity of object categorizations for the recognition of spatial environments might be sometimes a bit different from the granularity of categorizations for the object recognition. For example, in the task of recognizing spatial environment, books in the bookshelf are grouped as one entity: "some books". That is, for the configuration of the spatial environment, people will group all the books in the bookshelf, and treat them as a unit for the component of the configuration. On the other hand, people may use object parts as single objects. For example, people use sides of a room as single objects, e.g., floors, ceilings, walls, etc. It is a piece of future work on the granularity of objects that is suitable for the primitive components for cognitive tasks in vista spatial cognition.

7.3.2 Object Categories Recognized by Different Sensors

Blinded bats could fly, avoid obstacles, land on walls and ceiling, and survive in nature as well as bats with sight.

– L. Spallanzani

As visual creatures, humans have great difficulty imagining how animals that use sound, ultra-wave, smell, or magnetics orient in complex environments. At least nature tells us that vision is not the only possible way to "see" the world. Bats use sound as perceptive medium, rattle snakes use infrared ray as medium, pigeons use their magnetic sensors, ants use their noses, etc. It is not important what kinds of sensors are used, rather how to recognize the external world with the stimuli from the sensors. A sensor receives features of the external world from certain perspective. Different sensors can receive completely different features. Features obtained through visions are different from features obtained through noses. This leads to a research topic of *object recognition through a given sensors* and *preferred categories of the given sensors*: What is a possible taxonomy of categories through a sensor? Given a taxonomy acquired by certain sensor, what can be the possible taxonomy of categories? Which level of abstraction in the taxonomy is the level at which categories carry the most information, posses the highest cue validity, and are, thus, the most differentiated from one another?

7.3.3 Detecting Objects' Movements

In The Theory of Cognitive Prism object mapping is restricted to the mapped *fiat* containers of two spatial environments. After a spatial environment is recognized, cognitive agents can further address the task of movement detection, which is a general object mapping task between the two sequential snapshots of a spatial environment. To detect object's movement is to explore principles of object mapping between un-mapped *fiat* containers of two sequential snapshots of a spatial environment.

7.3.4 Humor: A Window to the Commonsense Knowledge

One of the mainstays of The Theory of Cognitive Prism is the commonsense knowledge of relatively stability, which is explored from commonsense understanding and evidenced by linguistic spatial descriptions. People have commonsense knowledge, but it is quite difficult for them to describe their commonsense knowledge explicitly, though they use it in their everyday life. People laugh, when they hear something against their commonsense. Therefore, it is possible to explore the commonsense knowledge through their laughter. This may link humor research with the commonsense research. For example, people laugh, when they hear *how do five Poles fix a lamp on the ceiling?– one stands on the table holding the lamp, and the four turning the table round and round.* We know that these people have the common sense that *to fix a lamp on the ceiling, one should stand on the table turning the lamp.*

7.3.5 Situations of Spatial Environments

The process of judgment on whether the perceived environment is the target is separated from the process of finding the compatibility between the perceived environment and the target one. Although in everyday life situation, the more compatible they are, the higher degree the perceived environment is believed as the target one, there are situations that such a direct relation might not be held. For example, if it is the earthquake situation, then the perceived environment still can be the target one, even there are spatial differences of rarely moved objects.

When we navigate in spatial environments, we see a sequence of spatial environments –from one spatial environment into the next one. When we go to our offices from home, we have already entered and left lots of spatial environments, e.g., the spatial environments of home, street, university, inside of office building, corridor, etc. In each spatial environment we believe that it is the target environment, so that we are already sure that it is our office, even before entering the room and look at it. In this case the spatial differences in the perceived environment provide information for situation detection—what happened in this environment? For example, if there are neither spatial differences of rarely moved objects, nor spatial differences of seldom moved objects, then the situation can be the everyday life situation. If there are more new objects in the perceived environment and all the remembered objects are there, it can be a moving-in situation. If no new objects appear in the perceived environment and some objects disappear in the remembered environment, it can be a moving-out situation. If there are spatial differences of locations of big furniture and there are neither new objects appearing nor remembered objects disappearing, it can be an after-a-party situation. If there are spatial differences of non-movable objects, it can even be an earthquake situation. A future work is to explore qualitative structure of situations.

References

Appleyard, D.: Why Buildings Are Known. Environment and Behaviour 1, 131–159 (1969)
Barkowsky, T.: Mental processing of geographic knowledge. In: Montello, D.R. (ed.) Spatial Information Theory - Foundations of Geographic Information Science, pp. 371–386. Springer, Berlin (2001)
Barkowsky, T.: Mental Representation and Processing of Geographic Knowledge: A Computational Approach. PhD thesis, Department for Informatics, University of Hamburg (2002)
Barkowsky, T.: Modeling mental spatial knowledge processing as integrating paradigm for spatial reasoning. In: Guesgen, H.W., Mitra, D., Renz, J. (eds.) Foundations and applications of spatio-temporal reasoning (FASTR) - Papers from the 2003 AAAI Spring Symposium (Technical Report SS-03-03), pp. 1–2. AAAI Press, Menlo Park (2003)
Barkowsky, T., Freksa, C., Hegarty, M., Lowe, R.: Reasoning with mental and external diagrams: computational modeling and spatial assistance. In: Papers from the 2005 AAAI Spring Symposium. Technical Report SS-05-06. AAAI Press, Menlo Park (2005)
Bennett, B.: The role of definitions in construction and analysis of formal ontologies. In: Sixth Symposium on Logical Formalizations of Commonsense Reasoning, AAAI Spring Symposium, Palo Alto (2003)
Bennett, B.: Relative definability in formal ontologies. In: Varzi, A., Vieu, L. (eds.) Proceedings of the 3rd International Conference on Formal Ontology in Information Systems (FOIS 2004), pp. 107–118. IOS Press, Amsterdam (2004)
Bennett, B., Cohn, A., Wolter, F., Zakharyaschev, M.: Multi-dimensional modal logic as a framework for spatio-temporal reasoning. Applied Intelligence 17(3), 239–251 (2002)
Bennett, B., Cohn, A.G., Torrini, P., Hazarika, S.M.: A foundation for region-based qualitative geometry. In: Horn, W. (ed.) Proceedings of ECAI 2000, pp. 204–208 (2000a)
Bennett, B., Cohn, A.G., Torrini, P., Hazarika, S.M.: Region-based Qualitative Geometry. Technical Report, School of Computing, University of Leeds (2000b)
Biederman, I.: Recognition-by-components: A Theory of Human Image Understanding. Psychological Review 94(2), 115–147 (1987)
Biederman, I., Gerhardstein, P.C.: Recognizing depth-rotated object priming. Journal of Experimental Psychology: Human Perception and Performance 19, 1162–1182 (1993)
Biederman, I., Gerhardstein, P.C.: Viewpoint-dependent mechanisms in visual object recognition: Reply to Tarr and Buelthoff. Journal of Experimental Psychology: Human Perception and Performance 21(6), 1506–1514 (1995)
Boulding, K.: The image: Knowledge in life and society. University of Michigan, Ann Arbor (1956)

Brennan, J., Martin, E., Kim, M.: Developing An Ontology Of Spatial Relations. In: Gero, J., Tversky, B., Knight, T. (eds.) Visual and Spatial Reasoning in Design III, pp. 163–182. MIT, Cambridge (2004)

Breuel, T.: Geometric Aspects of Visual Object Recognition. PhD thesis, MIT Artificial Intelligence Laboratory (1992)

Buelthoff, H.H., Edelman, S.: Psychophysical support for a two-dimensional view interpolation theory of object recognition. Proceedings of National Academy of Science, USA, 60–64 (1992)

Byrne, R.W.: Memory for urban geography. Quarterly Journal of Experimental Psychology 31, 147–154 (1979)

Cohen, R., Baldwin, L., Sherman, R.C.: Cognitive maps of naturalistic setting. Child Development 49, 11216–11218 (1978)

Cohn, A.G.: Modal and non modal qualitative spatial logics. In: Anger, F.D., Guesgen, H.M., van Benthem, J. (eds.) Proceedings of the Workshop on Spatial and Temporal Reasoning. IJCAI, Chambéry (1993)

Cohn, A.G.: A hierarchical representation of qualitative shape based on connection and convexity. In: Kuhn, W., Frank, A.U. (eds.) COSIT 1995. LNCS, vol. 988, pp. 311–326. Springer, Heidelberg (1995)

Cohn, A.G., Bennett, B., Gooday, J.M., Gotts, N.: RCC: a calculus for region based qualitative spatial reasoning. GeoInformatica 1, 275–316 (1997)

Cohn, A.G., Gotts, N.M.: The 'egg-yolk' representation of regions with indeterminate boundaries. In: Burrough, P., Frank, A.U. (eds.) Proceedings, GISDATA Specialist Meeting on Geographical Objects with Undetermined Boundaries, pp. 171–187. Francis & Taylor, Abington (1996)

Couclelis, H., Golledge, R.G.: and Gale, N., and Tobler, W. Exploring the anchor-point hypothesis of spatial cognition. Journal of Environmental Psychology 7, 99–122 (1987)

Creem, S.H., Proffitt, D.R.: Defining the cortical visual systems: "What", "Where", and "How". Acta Psychologica 107, 43–68 (2001)

Davis, E.: Organizing spatial knowledge. Department of Computer Science, Yale University, Research Rep. 193 (1981)

de Laguna, T.: Point, line and surface as sets of solids. The Journal of Philosophy 19, 449–461 (1922)

Dong, T.: SNAPVis and sPANVis: Ontologies for recognizing variable vista spatial environments. In: Freksa, C., Knauff, M., Krieg-Brückner, B., Nebel, B., Barkowsky, T. (eds.) Spatial Cognition IV. LNCS (LNAI), vol. 3343, pp. 344–365. Springer, Heidelberg (2005)

Downs, R., Stea, D.: Cognitive maps and spatial behaviour: Process and products. In: Buffart, E., Buffart, H. (eds.) Image and Environment, pp. 8–26. Aldine Publishing Co. (1973)

Edelman, S., Buelthoff, H.H.: Orientation dependence in the recognition of familiar and novel views of three-dimensional objects. Vision Research 32, 2385–2400 (1992)

Egenhofer, M.: A formal definition of binary topological relationships. In: Litwin, W., Schek, H.-J. (eds.) FODO 1989. LNCS, vol. 367, pp. 457–472. Springer, Heidelberg (1989)

Egenhofer, M.: Reasoning about binary topological relations. In: Günther, O., Schek, H.-J. (eds.) SSD 1991. LNCS, vol. 525, pp. 143–160. Springer, Heidelberg (1991)

Egenhofer, M.: A Model for Detailed Binary Topological Relationships. Geomatica 47(3 & 4), 261–273 (1993)

Egenhofer, M.: Deriving the Composition of Binary Topological Relations. Journal of Visual Languages and Computing 5(2), 133–149 (1994)

Egenhofer, M.: Spherical topological relations. Journal on Data Semantics III, 25–49 (2005)

Egenhofer, M., Franzosa, R.: Point-set topological spatial relations. International Journal of Geographical Information Systems 5, 161–174 (1991)

Egenhofer, M., Franzosa, R.: On the Equivalence of Topological Relations. International Journal of Geographical Information Systems 9(2), 133–152 (1995)

Egenhofer, M., Mark, D.: Naive Geography. In: Kuhn, W., Frank, A.U. (eds.) COSIT 1995. LNCS, vol. 988, pp. 1–15. Springer, Heidelberg (1995)

Egenhofer, M., Sharma, J.: Topological Relations Between Regions in R2 and Z2. In: Abel, D.J., Ooi, B.-C. (eds.) SSD 1993. LNCS, vol. 692. Springer, Heidelberg (1993)

Engel, D., Bertel, S., Barkowsky, T.: Spatial principles in control of focus in reasoning with mental representations, images, and diagrams. In: Freksa, C., Knauff, M., Krieg-Brückner, B., Nebel, B., Barkowsky, T. (eds.) Spatial Cognition IV. LNCS (LNAI), vol. 3343, pp. 181–203. Springer, Heidelberg (2005)

Eschenbach, C.: A mereotopological definition of "point". In: Eschenbach, C., Habel, C., Smith, B. (eds.) Topological Foundations of Cognitive Science, Papers from the Workshop at the First International Summer Institute in Cognitive Science, Buffalo, pp. 63–80 (1994); Reports of the Doctoral Program in Cognitive Science, No. 37

Foos, P.W.: Constructing Cognitive Maps From Sentences. Journal of Experimental Psychology: Human Learning and Memory 6(1), 25–38 (1980)

Franklin, N., Tversky, B.: Searching imagined environments. Journal of Experimental Psychology: General 119, 63–76 (1990)

Frege, G.: Die Grundlagen der Arithmetik. Köbner, Breslau (1884); cited according to the English translation by Austin, J. L.: Foundation of Arithmetic. Basil Blackwell, Oxford (1950)

Freksa, C.: Using Orientation Information for Qualitative Spatial Reasoning. In: Frank, A.U., Formentini, U., Campari, I. (eds.) GIS 1992. LNCS, vol. 639, Springer, Heidelberg (1992)

Gooday, J.M., Cohn, A.G.: Conceptual Neighbourhoods in Spatial and Temporal Reasoning. In: Proceedings ECAI 1994 Workshop on Spatial and Temporal Reasoning. Rodríguez, R (1994)

Gotts, N.M.: How far can we C? defining a 'doughnut' using connection alone. In: Doyle, J., Sandewall, E., Torasso, P. (eds.) Principles of Knowledge Representation and Reasoning: Proceedings of the 4th International Conference (KR 1994), Morgan Kaufmann, San Francisco (1994)

Goyal, R.: Similarity Assessment for Cardinal Directions between Extended Spatial Objects. PhD thesis, Spatial Information Science and Engineering, University of Maine (2000)

Grenon, P., Smith, B.: SNAP and SPAN: Towards Dynamic Spatial Ontology. In: Spatial Cognition and Computation, vol. 4(1), pp. 69–103. Lawrence Erlbaum Associates, Inc., Mahwah (2004)

Guesgen, H.: Spatial Reasoning Based on Allen's Temporal Logic. Technical Report TR-89-949, International Computer Science Institute, Berkeley, CA (1989)

Haar, R.: Computational Models of Spatial Relations. Technical Report TR-478, MSC-72-03610, Computer Science Department, University of Maryland, College Park, MD (1976)

Hardt, S.: Physics, Naive. In: Shapiro, S. (ed.) Encyclopedia of Artificial Intelligence, 2nd edn., pp. 1147–1149. John Wiley & Sons, Inc., New York (1992)

Hayes, P.J.: The naïve physics manifesto. In: Michie, D. (ed.) Expert Systems in the Micro-Electronic Age, pp. 242–270. Edinburgh University Press, Edinburgh (1978)

Hernández, D.: Qualitative Representation of Spatial Knowledge. Springer, Heidelberg (1994)

Hirtle, S.C., Jonides, J.: Evidence of hierarchies in cognitive maps. Memory & Cognition 13(3), 208–217 (1985)

Humphreys, G.K., Khan, S.C.: Recognizing novel views of three-dimensional objects. Canadian Journal of Psychology 46, 170–190 (1992)

Jolicoeur, P., Gluck, M.A., Kosslyn, S.M.: From pictures to words: Making the connection. Cognitive Psychology 16, 243–275 (1984)

Kaplan, S.: Cognitive maps in perception and thought. In: Downs, R.M., Stea, D. (eds.) Image and environment, pp. 63–78. Aldine, Chicago (1973)

Klippel, A.: Wayfinding choremes. In: Kuhn, W., Worboys, M.F., Timpf, S. (eds.) COSIT 2003. LNCS, vol. 2825, pp. 301–315. Springer, Heidelberg (2003a)

Klippel, A.: Wayfinding Choremes–Conceptualizing Wayfinding and Route Direction Elements. PhD thesis, Department of Math and Informatics, University of Bremen (2003b)

Klippel, A., Dewey, C., Knauff, M., Richter, K.-F., Montello, D.R., Freksa, C., Loeliger, E.-A.: Direction concepts in wayfinding assistance systems. In: Baus, J., Kray, C., Porzel, R. (eds.) Workshop on Artificial Intelligence in Mobile Systems (AIMS 2004), pp. 1–8 (2004); SFB 378 Memo 84; Saarbrücken

Klippel, A., Lee, P.U., Fabrikant, S., Montello, D.R., Bateman, J.: The cognitive conceptual approach as a leitmotif for map design. In: AAAI 2005 Spring Symposium on Reasoning with Mental and External Diagrams: Computational Modeling and Spatial Assistance (2005a)

Klippel, A., Montello, D.R.: On the robustness of mental conceptualizations of turn direction concepts. In: Egenhofer, M.J., Freksa, C., Miller, H. (eds.) GIScience 2004. The Third International Conference on Geographic Information Science, pp. 139–141 (2004); Regents of the University of California; Maryland, USA. Extended Abstract

Klippel, A., Richter, K.-F.: Chorematic focus maps. In: Gartner, G. (ed.) Location Based Services & Telecartography, pp. 39–44. Technische Universität Wien, Wien (2004)

Klippel, A., Richter, K.-F., Barkowsky, T., Freksa, C.: The cognitive reality of schematic maps. In: Meng, L., Zipf, A., Reichenbacher, T. (eds.) Map-based Mobile Services - Theories, Methods and Implementations, pp. 57–74. Springer, Berlin (2005b)

Klippel, A., Tappe, H., Kulik, L., Lee, P.U.: Wayfinding choremes - a language for modeling conceptual route knowledge. In: Journal of Visual Languages and Computing (2005c)

Koehler, W.: Gestalt Psychology. Liveright, London (1929)

Koffka, K.: Principles of Gestalt Psychology. Brace & World. Harcourt, New York (1935)

Kosslyn, S.M., Pick, H.L., Fariello, G.R.: Cognitive maps in children and men. Child Development 45, 707–716 (1974)

Kuipers, B.: A frame for frames: Representing Knowledge for Recognition. Technical report, MIT AI, Memo 322 (1975)

Kuipers, B.: Representing Knowledge of Large-Scale Space. PhD thesis, Mathematics Department, Massachusetts Institute of Technology, Cambridge, Massachusetts (1977)

Kuipers, B.: Modeling spatial knowledge. Cognitive Science 2, 129–153 (1978)

Kuipers, B.: On Representing Commonsense Knowledge. In: Findler, N.V. (ed.) Associative Networks: The Representation and Use of Knowledge by Computers, pp. 393–408. Academic Press, NY (1979)

Leach, E.: Anthropological aspects of language: Animal categories and verbal abuse. In: Lenneberg, E.H. (ed.) New directions in the study of language. MIT Press, Cambridge (1964)

Lee, P.U.: Costs of Switching Perspectives in Route and Survey Descriptions. PhD thesis, Department of Psychology, Stanford University (2002)

Lee, T.: Urban Neighbourhood as a Socio-Spatial Schema. Human Relations 21, 241–267 (1968)

Liter, J.C., Buelthoff, H.H.: An Introduction to Object Recognition. Technical report, Max-Planck-Institue fur biologische Kybernetik. Technical Report No. 43 (1996)
Lynch, K.: The Image of The City. The MIT Press, Cambridge (1960)
Marr, D.: Vision: a computational investigation into the human representation and processing of visual information. Freeman, San Francisco (1982)
Marr, D., Nishihara, H.K.: Representation and recognition of the spatial organization of three dimensional shapes. Proceedings of the Royal Society of London B 200 200, 269–294 (1978)
McDermott, D.: Finding objects with given spatial properties. Technical report, Department of Computer Science, Yale University. Research Rep. 195 (1981)
McKeithem, K.B., Reitman, J.S., Rueter, H.R., Hirtle, S.C.: Knowledge organization and skill differences in computer programmers. Cognitive Psychology 13, 307–325 (1981)
McNamara, T., Ratcliff, R., McKoon, G.: The mental representation of knowledge acquired from maps. Journal of Experimental Psychology: Learning, Memory, and Cognition 10, 723–732 (1984)
McNamara, T.P.: Mental Representation of Spatial Relations. Cognitive Psychology 18, 87–121 (1986)
McNamara, T.P.: Memory's View of Space. The Psychology of Learning and Motivation 27, 147–186 (1991)
McNamara, T.P.: Spatial Representation. Geoforum 23(2), 139–150 (1992)
Minsky, M.: A Framework for Representing Knowledge. In: Winston, P.H. (ed.) The Psychology of Computer Vision. McGraw-Hill, New York (1975)
Montello, D.: Scale and Multiple Psychologies of Space. In: Frank, A., Campari, I. (eds.) Spatial information theory: A theoretical basis for GIS, pp. 312–321. Springer, Berlin (1993)
Murphy, G.L., Smith, E.E.: Basic-level superiority in picture categorization. Journal of Verbal Learning and Verbal Behavior 21, 1–20 (1982)
Newcombe, N., Liben, L.: Barrier effects in cognitive maps of children and adults. Journal of Experimental Child Psychology 34, 46–58 (1982)
Newell, A.: Physical symbol systems. Cognitive Science 4, 135–183 (1980)
Newell, A., Simon, H.A.: Human problem solving. Prentice Hall, Englewood Cliffs (1972)
Palmer, S., Rosch, E., Chase, P.: Canonical perspective and the perception of objects. In: Long, J., Baddeley, A. (eds.) Attention & performance IX, pp. 135–151. Lawrence Erlbaum, Mahwah (1981)
Piaget, J.: The Construction of Reality in the Child. Routledge & Kegan Paul Ltd, London (1954)
Piaget, J., Inhelder, B.: La représentation de l'espace chez l'enfant. In: Bibliothèque de Philosophie Contemporaine, PUF, Paris (1948); English translation by Langdon, F. J., Lunzer, J. L. (1956)
Randell, D., Cui, Z., Cohn, A.: A spatial logic based on regions and connection. In: Nebel, B., Swartout, W., Rich, C. (eds.) Proc. 3rd Int. Conf. on Knowledge Representation and Reasoning, pp. 165–176. Morgan Kaufmann, San Mateo (1992)
Rock, I., Palmer, S.: The Legacy of Gestalt Psychology. Scientific American, 48–61 (1990)
Rosch, E.: Cognitive Reference Points. Cognitive Psychology 7(4), 532–547 (1975)
Rosch, E., Mervis, C.B., Gray, W., Johnson, D., Boyes-Braem, P.: Basic objects in natural categories. Cognitive Psychology 8, 382–439 (1976)
Rueckl, J., Cave, K., Kosslyn, S.: Why are "What" and "Where" Processed by Separate Cortical Visual Systems? A Computational Investigation. Journal of Cognitive Neuroscience 1, 171–186 (1989)

Sadalla, E., Burroughs, W.J., Staplin, L.J.: Reference points in spatial cognition. Journal of Experimental Psychology: Human Learning and Memory 6(5), 516–528 (1980)

Sadalla, E., Magel, S.: The Perception of Traversed Distance. Environment and Behavior 12(2), 65–79 (1980)

Schlieder, C.: Reasoning about Ordering. In: Kuhn, W., Frank, A.U. (eds.) COSIT 1995. LNCS, vol. 988. Springer, Heidelberg (1995)

Schmidtke, H.R.: The house is north of the river: Relative localization of extended objects. In: Montello, D.R. (ed.) COSIT 2001. LNCS, vol. 2205, pp. 415–430. Springer, Heidelberg (2001)

Shemyakin, F.N.: General problems of orientation in space and space representations. In: Anayev, B.G. (ed.) Psychological Science in the USSR, Washington, D.C. U.S. Office of Technical Reports, pp. 186–251 (1962)

Siegel, A.W., White, S.H.: The development of spatial representation of large-scale environments. In: Reese, H. (ed.) Advances in child development and behaviour, pp. 9–55. Academic Press, San Diego (1975)

Smith, B.: Topological Foundations of Cognitive Science. In: Eschenbach, C., Habel, C., Smith, B. (eds.) Topological Foundations of Cognitive Science. Buffalo, NY (1994); Workshop at the FISI-CS

Smith, B.: The Structure of the Common-Sense World. Acta Philosophica Fennica 58, 290–317 (1995)

Smith, B.: Mereotopology: A Theory of Parts and Boundaries. Data and Knowledge Engineering 20, 287–303 (1996)

Smith, B.: Fiat objects. Topoi 20(2), 131–148 (2001)

Smith, B., Varzi, A.C.: Fiat and bona fide boundaries. Philosophy and Phenomenological Research 60(2), 401–420 (2000)

Spelke, E.S.: Principles of Object Perception. Cognitive Science 14, 29–56 (1990)

Stevens, A., Coupe, P.: Distance estimation from cognitive maps. Cognitive Psychology 13, 526–550 (1978)

Talmy, L.: How Language Structures Space. In: Pick, H., Acredolo, L. (eds.) Spatial Orientation: Theory, Research and Application, pp. 225–281. Plenum Press, New York (1983)

Tarr, M.: Rotating objects to recognizing them: a case study of the role of viewpoint dependency in the recognition of three-dimensional objects. Psychonomic Bulletin and Review 2(1), 55–82 (1995)

Tarr, M.J., Buelthoff, H.H.: Is Human Object Recognition Better Described by Geon Structural Descriptions or by Multiple Views? (1995); Comment on Biederman and Gerhardstein. Journal of Experimental Psychology: Human Perception and Performance 21(6),1494–1505 (1995)

Tarr, M.J., Buelthoff, H.H.: Image-based object recognition in man, monkey and machine. Cognition 67, 1–20 (1998)

Thorndyke, P.: Distance estimeation from cognitive maps. Cognitive Psychology 13, 526–550 (1981)

Tolman, E.C.: Cognitive Maps in Rats and Men. The Psychological Review 55(4), 189–208 (1948)

Trowbridge, C.: On fundamental methods of orientation and 'imaginary maps'. Science 88, 888–897 (1933)

Tversky, B.: Distortions in Memory for Maps. Cognitive Psychology 13, 407–433 (1981)

Tversky, B.: Spatial Mental Models. The Psychology of Learning and Motivation 27, 109–145 (1991)

Tversky, B.: Distortions in Cognitive Maps. Geoforum 23(2), 131–138 (1992)

Tversky, B.: Cognitive maps, cognitive collages, and spatial mental models. In: Frank, A., Campari, I. (eds.) Spatial information theory — A theoretical basis for GIS, pp. 14–24. Springer, Heidelberg (1993)

Tversky, B.: Functional significance of visuospatial representations. In: Shah, P., Miyake, A. (eds.) Handbook of higher-level visuospatial thinking. Cambridge University Press, Cambridge (2005)

Tversky, B., Lee, P.U.: How space structures language. In: Freksa, C., Habel, C., Wender, K.F. (eds.) Spatial Cognition 1998. LNCS (LNAI), vol. 1404, pp. 157–176. Springer, Heidelberg (1998)

Tversky, B., Lee, P., Mainwaring, S.: Why do speakers mix perspectives. Spatial Cognition and Computation 1, 399–412 (1999a)

Tversky, B., Morrison, J.B., Franklin, N., Bryant, D.: Three spaces of spatial cognition. Professional Geographer 51, 516–524 (1999b)

Ullmer-Ehrich, V.: The Structure of Living Space Descriptions. In: Jarvella, R.J., Klein, W. (eds.) Speech, Place, and Action, pp. 219–249. John Wiley & Sons Ltd, Chichester (1982)

Ungerleider, L.G., Mishkin, M.: Two cortical visual systems. In: Ingle, D.J., Goodale, M.A., Mansfield, R.J.W. (eds.) Analysis of visual behavior, pp. 157–165. The MIT Press, Cambridge (1982)

Wertheimer, M.: Numbers and numerical concepts in primitive peoples. In: Ellis, W.D. (ed.) A Source Book of Gestalt Psychology. Brace Co., Harcourt, New York (1938)

Wertheimer, M.: Principles of perceptual organization. In: Beardsle, D., Wertheimer, M. (eds.) Reading in Perception. Van Nostrand, New York (1958)

Wilson, B., Baddeley, A., Young, A.: LE, A Person Who Lost Her 'Mind's Eye'. Neurocase 5, 119–127 (1999)

Woodcock, J., Davies, J.: Using Z: Specification, Refinement and Proof. Prentice-Hall, Englewood Cliffs (1996)

Yeap, W.: Towards a computational theory of cognitive maps. Artificial Intelligence 34, 297–360 (1988)

Yeap, W., Jefferies, M.E.: Computing a representation of the local environment. Artificial Intelligence 107, 265–301 (1999)

Zimmermann, K.: Enhancing Qualitative Spatial Reasoning- Combining Orientation and Distance. In: Campari, I., Frank, A.U. (eds.) COSIT 1993. LNCS, vol. 716. Springer, Heidelberg (1993)

Appendix A
Relations between the Sizes of Regions

When you sit on the couch, you can see the books on the writing-desk, although you cannot reach them by your arms. This implies that the light beam which is connected with the books and your eyes has larger size than your arm which cannot be connected with the books and the eyes. Sizes of regions can be compared as follows: Given two regions A and B, if region X_1 can be moved to such a place that it is connected with both A and B, while for region X_0, there is no such a place, then the size of X_0 is smaller than the size of X_1, de Laguna (1922), shown in Figure A.1. In the notion of category, sizes of regions are compared as follows: Given two regions A and B, if there is a region of the same category as region X_1 which is connected with both A and B, while for region X_0, there is no region of the same category as X_0 which is connected with A and B, then the size of X_0 is smaller than the size of X_1. Two regions X_0 and X_1 of the same size is explained that for any two regions A and B, if both X_0 and X_1 can be connected with A and B or neither of X_0 and X_1 can be connected with A and B, shown in Figure A.2. In the notion of category, two regions being of the same size is explained as follows: for any two regions A and B, if there is a region of the same category as region X_1 which is connected with both A and B, then there is a region of the same category as region X_0 which is connected with both A and B, and if there is a region of the same category as region X_0 which is connected with both A and B, then there is a region of the same category as region X_1 which is connected with both A and B, then the size of X_0 is the same as the size of X_1. The basic relations of sizes are *smaller than* (symbolically represented by '$<_s$', '$_s$' represents 'size') and *equal* (symbolically represented by '$=_s$').

Definition A.0.1. *Given two regions X_0 and X_1, then "X_0 is smaller than X_1" is defined as that there are two regions A and B such that the near extension of A by X_0 is disconnected with B and the near extension of A by X_1 is connected with B.*

$$X_0 <_s X_1 \stackrel{\text{def}}{=} \exists A \exists B \bullet \neg \mathbf{C}(A^{X_0}, B) \wedge \mathbf{C}(A^{X_1}, B)$$

Definition A.0.2. *Given two regions X_0 and X_1, then "X_0 is of the same size as X_1" is defined as all regions A and B, it holds that either the near extensions of A by X_0 and X_1 are connected with both A and B, or the near extensions of A by X_0 and X_1 are not connected with both A and B.*

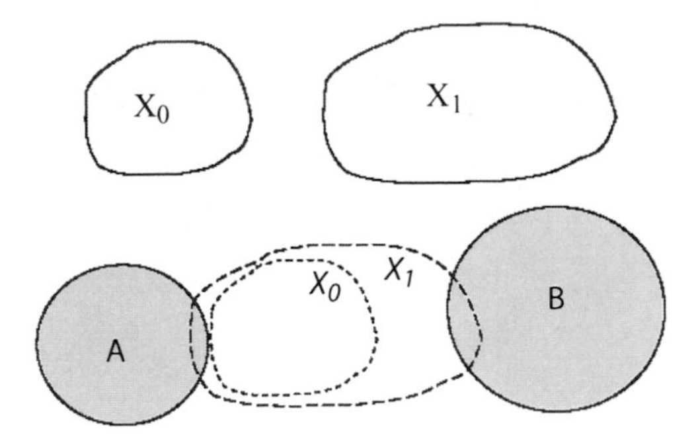

Fig. A.1 The fact that region X_0 is smaller than region X_1 can be tested by given two regions A and B as follows: If X_1 can be moved to such a place that it is connected both with A and B, while for X_0 there is no such a place

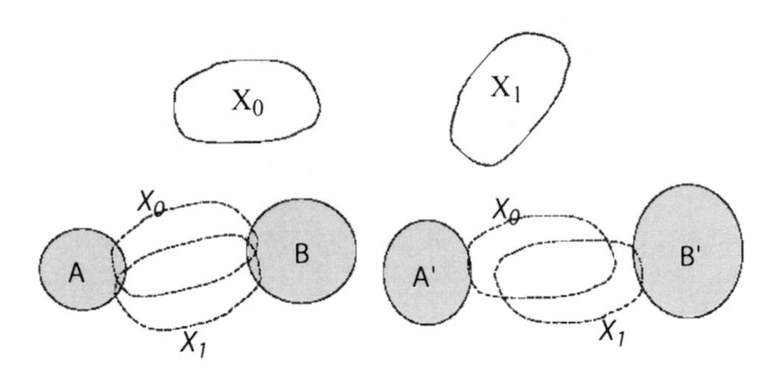

Fig. A.2 The fact that region X_0 is of the same size as region X_1 can be tested as follows: For any two regions A and B, if X_1 can be moved to such a place that it is connected both with A and B, so can X_0, and there is no such a place for X_1 that it is connected both with A and B, neither is X_0

$$X_0 =_s X_1 \stackrel{\text{def}}{=} \forall A, B \bullet \mathbf{C}(A^{X_0}, B) \wedge \mathbf{C}(A^{X_1}, B)$$
$$\vee \neg \mathbf{C}(A^{X_0}, B) \wedge \neg \mathbf{C}(A^{X_1}, B)$$

Theorem A.0.1. *Given two regions, X_0 (of category $\mathtt{X}_0$) and X_1 (of category $\mathtt{X}_1$), all regions A and B, it holds that if $X_0 <_s X_1$, then if A^{X_0} is connected with B, then A^{X_1} is also connected with B.*

$$\forall A, B \bullet X_0 <_s X_1 \rightarrow (\mathbf{C}(A^{X_0}, B) \rightarrow \mathbf{C}(A^{X_1}, B))$$

Theorem A.0.2. *For any regions X_0, X_1, X_2, if $X_0 <_s X_1$ and $X_1 <_s X_2$, then $X_0 <_s X_2$.*

$$\forall X_0, X_1, X_2 \bullet X_0 <_s X_1 \wedge X_1 <_s X_2 \rightarrow X_0 <_s X_2$$

Theorem A.0.3. *For any region X, X is of the same size as itself.*

$$\forall X \bullet X =_s X$$

Theorem A.0.4. *For all regions X_0 and X_1, it holds that either X_0 is smaller than X_1, or X_0 equals to X_1, or X_1 is smaller than X_0.*

$$\forall X_0, X_1 \bullet X_0 <_s X_1 \vee X_0 =_s X_1 \vee X_1 <_s X_0$$

Proofs of the above theorems are listed in Appendix B.

Appendix B
Theorem Proof Sketches

Theorem 5.2.1. Given A an object region, and $\mathtt{X}$ be a category of an object region. There is X satisfying that X is a member of $\mathtt{X}$ such that if X is connected with A, then there is Y such that all W, it holds that Y is connected with W, if and only if there is V satisfying that V is a member of $\mathtt{X}$ and connected with A such that V is connected with W.

$$\exists X | X \in \mathtt{X} \bullet \mathbf{C}(X,A) \rightarrow \exists Y \forall W \bullet (\mathbf{C}(W,Y) \equiv \exists V | V \in \mathtt{X} \wedge \mathbf{C}(A,V) \bullet \mathbf{C}(W,V)))$$

Proof sketch:

$(1)[definition\ of\ \equiv]$
$\forall W \bullet \mathbf{C}(W,X) \equiv \mathbf{C}(W,X)$
$(2)[$**Axiom 5.2.3**$]$
$\exists X | X \in \mathtt{X} \bullet \mathbf{C}(X,A) \rightarrow \mathbf{C}(A,X)$
$(3)[(1),\ (2)]$
$\exists X | X \in \mathtt{X} \bullet \mathbf{C}(X,A) \rightarrow \forall W \bullet (\mathbf{C}(W,X) \equiv \mathbf{C}(A,X) \wedge \mathbf{C}(W,X))$
$(4)[\exists\ - introduce]$
$\exists X | X \in \mathtt{X} \bullet \mathbf{C}(X,A) \rightarrow \exists Y \forall W \bullet (\mathbf{C}(W,Y) \equiv \exists V \bullet V \in \mathtt{X} \wedge \mathbf{C}(A,V) \wedge \mathbf{C}(W,V))))$
$(5)[\,|\ - introduce]$
$\exists X | X \in \mathtt{X} \bullet \mathbf{C}(X,A) \rightarrow \exists Y \forall W \bullet (\mathbf{C}(W,Y) \equiv \exists V | V \in \mathtt{X} \wedge \mathbf{C}(A,V) \bullet \mathbf{C}(W,V)))$

The existence of A^X
Proof sketch:

$(1)[$**Theorem 5.2.1**$]$
$\exists X | X \in \mathtt{X} \bullet \mathbf{C}(X,A) \rightarrow \exists Y \forall W \bullet (\mathbf{C}(W,Y) \equiv \exists V | V \in \mathtt{X} \wedge \mathbf{C}(A,V) \bullet \mathbf{C}(W,V)))$
$(2)[$**Axiom 5.2.1**$]$
$\forall A \bullet \mathbf{C}(A,A) \rightarrow \forall \mathtt{Z} \exists Z | Z \in \mathtt{Z} \bullet \mathbf{C}(A,Z) \wedge \mathbf{C}(A,Z)$

$$(3)[(2), \textbf{Axiom 5.2.2}]$$
$$\forall \mathtt{Z} \exists Z | Z \in \mathtt{Z} \bullet \mathbf{C}(A,Z) \wedge \mathbf{C}(A,Z)$$
$$(4)[(3), \forall - elimination,\ p \wedge p \equiv p]$$
$$\exists X | X \in \mathtt{X} \bullet \mathbf{C}(A,X)$$
$$(5)[(1),(4), \textbf{Axiom 5.2.3}]$$
$$\exists Y \forall W \bullet (\mathbf{C}(W,Y) \equiv \exists V | V \in \mathtt{X} \wedge \mathbf{C}(A,V) \bullet \mathbf{C}(W,V)))$$

The uniqueness of A^X
Proof sketch:

$$(1)[\textbf{Definition 5.2.2}, \textbf{Theorem 5.2.1}]$$
$$\forall W \bullet (\mathbf{C}(W,Y_1) \equiv \exists V | V \in \mathtt{X} \wedge \mathbf{C}(A,V) \bullet \mathbf{C}(W,V))$$
$$(2)[\textbf{Definition 5.2.2}, \textbf{Theorem 5.2.1}]$$
$$\forall W \bullet (\mathbf{C}(W,Y_2) \equiv \exists V | V \in \mathtt{X} \wedge \mathbf{C}(A,V) \bullet \mathbf{C}(W,V))$$
$$(3)[(1),(2), definition\ of\ \equiv]$$
$$\forall W \bullet \mathbf{C}(W,Y_1) \equiv \mathbf{C}(W,Y_2)$$
$$(4)[(3), \textbf{Definition 5.2.1}]$$
$$\mathbf{P}(Y_1,Y_2) \wedge \mathbf{P}(Y_2,Y_1)$$
$$(5)[(4), \textbf{Axiom 5.2.4}]$$
$$Y_1 = Y_2$$

Theorem 5.2.2 Given object regions A and X, A is a part of the *near* extension of A by X.

$$\forall A, X \bullet \mathbf{P}(A, A^X)$$

Proof sketch:

$$(1)[\textbf{Axiom 5.1}]$$
$$\forall A, B \bullet \mathbf{C}(A,B) \rightarrow \forall \mathtt{Z} \exists Z | Z \in \mathtt{Z} \bullet \mathbf{C}(A,Z) \wedge \mathbf{C}(B,Z)$$
$$(2)[(1), \forall - elimination]$$
$$\forall A, B \bullet \mathbf{C}(A,B) \rightarrow \exists Z | Z \in \mathtt{X} \bullet \mathbf{C}(A,Z) \wedge \mathbf{C}(B,Z)$$
$$(3)[\textbf{Definition 5.2.2}\ of\ A^X]$$
$$\forall W \bullet \mathbf{C}(W,A^X) \equiv \exists Z | Z \in \mathtt{X} \wedge \mathbf{C}(A,Z) \bullet \mathbf{C}(W,Z)$$
$$(4)[(3), \forall - elimination,\ \exists x | p \bullet q \equiv \exists x \bullet p \wedge q]$$
$$\mathbf{C}(B,A^X) \equiv \exists Z | Z \in \mathtt{X} \bullet \mathbf{C}(A,Z) \wedge \mathbf{C}(B,Z)$$
$$(5)[(2),(4)]$$
$$\forall A, B \bullet \mathbf{C}(A,B) \rightarrow \mathbf{C}(B,A^X)$$

$(6)[(5),\ \textbf{Axiom 5.2.3}]$
$\forall A,B \bullet \mathbf{C}(B,A) \to \mathbf{C}(B,A^X)$
$(7)[\textbf{Definition}\ of\ \mathbf{P}\ in\ \textbf{Definition 5.2.1}]$
$\forall A,X \bullet \mathbf{P}(A,A^X)$

Theorem 5.2.3. All object regions A, B and X, it holds that the *near* extension of A by X is connected with B, if and only if A is connected with the *near* extension of B by X.

$$\forall A,B,X \bullet \mathbf{C}(A^X,B) \equiv \mathbf{C}(A,B^X)$$

Proof sketch:

$(1)[\textbf{Axiom 5.2.3}]$
$\forall A,B,X \bullet \mathbf{C}(A^X,B) \to \mathbf{C}(B,A^X)$
$(2)[\textbf{Axiom 5.2.3}]$
$\forall A,B,X \bullet \mathbf{C}(B,A^X) \to \mathbf{C}(A^X,B)$
$(3)[(1),(2),\textbf{Definition}\ of\ \equiv]$
$\forall A,B,X \bullet \mathbf{C}(A^X,B) \equiv \mathbf{C}(B,A^X)$
$(4)[\textbf{Definition 5.2.2}\ of\ A^X,\ \exists x|p \bullet q \equiv \exists x \bullet p \wedge q]$
$\mathbf{C}(B,A^X) \stackrel{\text{def}}{=} \exists V|V \in \mathtt{X} \bullet \mathbf{C}(A,V) \wedge \mathbf{C}(B,V)$
$(5)[p \wedge q \equiv q \wedge p]$
$\exists V|V \in \mathtt{X} \bullet \mathbf{C}(A,V) \wedge \mathbf{C}(B,V) \equiv \mathbf{C}(B,V) \wedge \mathbf{C}(A,V)$
$(6)[\textbf{Definition 5.2.2}\ of\ B^X,\ \exists x|p \bullet q \equiv \exists x \bullet p \wedge q]$
$\mathbf{C}(A,B^X) \stackrel{\text{def}}{=} \exists V|V \in \mathtt{X} \bullet \mathbf{C}(B,V) \wedge \mathbf{C}(A,V)$
$(7)[(3),(4),(5),(6)]$
$\forall A,B,X \bullet \mathbf{C}(A^X,B) \equiv \mathbf{C}(A,B^X)$

Theorem 5.2.4. Let C be a constructed region, and $\mathtt{X}$ be a category of an object region. There is X satisfying that X is a member of $\mathtt{X}$ such that if X is connected with C, there is Y such that all W, it holds that Y is connected with W, if and only if there is V satisfying that V is a member of $\mathtt{X}$ and connected with C such that V is connected with W.

$$\exists X|X \in \mathtt{X} \bullet \mathbf{C}(X,C) \to \exists Y \forall W \bullet (\mathbf{C}(W,Y) \equiv \exists V|V \in \mathtt{X} \wedge \mathbf{C}(C,V) \bullet \mathbf{C}(W,V)))$$

Proof sketch: similar with the proof sketch of **Theorem 5.2.1**.

The proofs of the existence and the uniqueness of C^X are similar to the proofs of those of A^X.

Theorem A.0.1. Given two regions, X_0 (of category $\mathtt{X}_0$) and X_1 (of category $\mathtt{X}_1$), all regions A and B, it holds that if $X_0 <_s X_1$, then if A^{X_0} is connected with B, then A^{X_1} is also connected with B.

$$\forall A, B \bullet X_0 <_s X_1 \rightarrow (\mathbf{C}(A^{X_0}, B) \rightarrow \mathbf{C}(A^{X_1}, B))$$

Proof sketch:

$$(1)\ [p \vee \neg p \equiv \mathtt{true}]$$
$$\forall A, B \bullet \mathbf{C}(A^{X_0}, B) \vee \neg \mathbf{C}(A^{X_0}, B) \vee \mathbf{C}(A^{X_1}, B) \vee \neg \mathbf{C}(A^{X_1}, B)$$
$$(2)\ [(1), p \vee q \equiv q \vee p]$$
$$\forall A, B \bullet (\mathbf{C}(A^{X_0}, B) \vee \neg \mathbf{C}(A^{X_1}, B)) \vee (\neg \mathbf{C}(A^{X_0}, B) \vee \mathbf{C}(A^{X_1}, B))$$
$$(3)\ [\neg p \vee q \equiv \neg(p \wedge \neg q), \neg p \vee q \equiv p \rightarrow q]$$
$$\forall A, B \bullet (\neg(\neg \mathbf{C}(A^{X_0}, B) \wedge \mathbf{C}(A^{X_1}, B))) \vee (\mathbf{C}(A^{X_0}, B) \rightarrow \mathbf{C}(A^{X_1}, B))$$
$$(4)\ [\textbf{Definition A.0.1}\ of\ <_s]$$
$$\forall A, B \bullet \neg(X_0 <_s X_1) \vee (\mathbf{C}(A^{X_0}, B) \rightarrow \mathbf{C}(A^{X_1}, B))$$
$$(5)\ [(4), \neg p \vee q \equiv p \rightarrow q]$$
$$\forall A, B \bullet (X_0 <_s X_1 \rightarrow (\mathbf{C}(A^{X_0}, B) \rightarrow \mathbf{C}(A^{X_1}, B)))$$

Theorem A.0.2. For any regions X_0, X_1, X_2, if $X_0 <_s X_1$ and $X_1 <_s X_2$, then $X_0 <_s X_2$.

$$\forall X_0, X_1, X_2 \bullet (X_0 <_s X_1 \wedge X_1 <_s X_2 \rightarrow X_0 <_s X_2)$$

Proof sketch:

$$(1)[\textbf{Definition A.0.1}\ of\ <_s]$$
$$\forall X_0, X_1, X_2 \bullet (X_0 <_s X_1 \wedge X_1 <_s X_2)$$
$$\stackrel{\text{def}}{=} \forall X_0, X_1, X_2 \bullet ((\exists A \exists B \bullet \neg \mathbf{C}(A^{X_0}, B) \wedge \mathbf{C}(A^{X_1}, B)) \wedge X_1 <_s X_2)$$
$$(2)\ [(1), ((\exists x \bullet p(x)) \wedge p) \rightarrow (\exists x \bullet (p(x) \wedge p))]$$
$$\forall X_0, X_1, X_2 \bullet (\exists A \exists B \bullet \neg \mathbf{C}(A^{X_0}, B) \wedge \mathbf{C}(A^{X_1}, B) \wedge X_1 <_s X_2)$$
$$(3)\ [(2), \exists - elimination]$$
$$\forall X_0, X_1, X_2 \bullet (\neg \mathbf{C}(A_0^{X_0}, B_0) \wedge \mathbf{C}(A_0^{X_1}, B_0) \wedge X_1 <_s X_2)$$
$$(4)\ [(3), p \wedge q \rightarrow q]$$
$$\forall X_0, X_1, X_2 \bullet (\mathbf{C}(A_0^{X_1}, B_0) \wedge X_1 <_s X_2)$$
$$(5)\ [(4), p \wedge q \rightarrow p]$$
$$\forall X_0, X_1, X_2 \bullet \mathbf{C}(A_0^{X_1}, B_0)$$
$$(6)\ [(4), p \wedge q \rightarrow q]$$
$$\forall X_0, X_1, X_2 \bullet X_1 <_s X_2$$

(7) $[(5),(6),\textbf{Theorem A.0.1}]$
$\mathbf{C}(A_0^{X_2}, B_0)$
(8) $[(3), p \land q \rightarrow p]$
$\forall X_0, X_1, X_2 \bullet \neg \mathbf{C}(A_0^{X_0}, B_0)$
(9) $[(3),(7),(8)]$
$\forall X_0, X_1, X_2 \bullet (\exists A_0 \exists B_0 \bullet \neg \mathbf{C}(A_0^{X_0}, B_0) \land \mathbf{C}(A_0^{X_2}, B_0))$
(10) $[(9), \textbf{Definition A.0.1}\ of\ <_s]$
$\forall X_0, X_1, X_2 \bullet (X_0 <_s X_2)$

Theorem A.0.3. For any region X, X is of the same size as itself.

$$\forall X \bullet X =_s X$$

Proof sketch:

(1) $[\textbf{Definition A.0.2}\ of\ =_s]$
$\forall X \bullet X =_s X$
$\stackrel{\text{def}}{=} \forall A, B \bullet ((\mathbf{C}(A^X, B) \land \mathbf{C}(A^X, B)) \lor (\neg \mathbf{C}(A^X, B) \land \neg \mathbf{C}(A^X, B)))$
(2) $[(1), p \land p \equiv p]$
$\forall A, B \bullet (\mathbf{C}(A^X, B) \lor \neg \mathbf{C}(A^X, B))$
(3) $[(2), p \lor \neg p \equiv \texttt{true}]$
`true`

Theorem A.0.4. For all regions X_0 and X_1, it holds that either X_0 is smaller than X_1, or X_0 equals to X_1, or X_1 is smaller than X_0.

$$\forall X_0, X_1 \bullet X_0 <_s X_1 \lor X_0 =_s X_1 \lor X_1 <_s X_0$$

Proof sketch:

(1) $[Suppose\ \neg(X_0 <_s X_1)]$
$\neg(X_0 <_s X_1)$
(2) $[\textbf{Definition A.0.1}\ of\ <_s]$
$\neg(\exists A \exists B \bullet \neg \mathbf{C}(A^{X_0}, B) \land \mathbf{C}(A^{X_1}, B))$
(3)$[(2), (\neg \exists x \exists y \bullet p \rightarrow \forall x \forall y \bullet \neg p)]$
$\forall A \forall B \bullet (\mathbf{C}(A^{X_0}, B) \lor \neg \mathbf{C}(A^{X_1}, B))$
(4) $[(3), p \equiv p \land (q \lor \neg q)]$
$\forall A \forall B \bullet ((\mathbf{C}(A^{X_0}, B) \lor \neg \mathbf{C}(A^{X_1}, B)) \land (\mathbf{C}(A^{X_0}, B) \lor \neg \mathbf{C}(A^{X_0}, B)))$

(5) $[(4), (p \vee q) \wedge (s \vee t) \equiv (p \wedge s) \vee (p \wedge t) \vee (q \wedge s) \vee (q \wedge t)]$

$\forall A \forall B \bullet \mathbf{C}(A^{X_0}, B) \wedge \mathbf{C}(A^{X_0}, B) \vee \neg\mathbf{C}(A^{X_1}, B) \wedge \mathbf{C}(A^{X_0}, B)$
$\quad \vee \mathbf{C}(A^{X_0}, B) \wedge \neg\mathbf{C}(A^{X_0}, B) \vee \neg\mathbf{C}(A^{X_1}, B) \wedge \neg\mathbf{C}(A^{X_0}, B)$

(6) $[(5), (p \wedge \neg p) \equiv \texttt{false}; p \wedge p \equiv p]$

$\forall A \forall B \bullet \mathbf{C}(A^{X_0}, B) \vee \neg\mathbf{C}(A^{X_1}, B) \wedge \mathbf{C}(A^{X_0}, B)$
$\quad \vee \texttt{false} \vee \neg\mathbf{C}(A^{X_1}, B) \wedge \neg\mathbf{C}(A^{X_0}, B)$

(7) $[(6), p \vee \texttt{false} \equiv p]$

$\forall A \forall B \bullet \mathbf{C}(A^{X_0}, B) \vee \neg\mathbf{C}(A^{X_1}, B) \wedge \mathbf{C}(A^{X_0}, B) \vee \neg\mathbf{C}(A^{X_1}, B) \wedge \neg\mathbf{C}(A^{X_0}, B)$

(8) $[(7), p \equiv p \wedge (q \vee \neg q)]$

$\forall A \forall B \bullet (\mathbf{C}(A^{X_0}, B) \vee \neg\mathbf{C}(A^{X_1}, B) \wedge \mathbf{C}(A^{X_0}, B) \vee \neg\mathbf{C}(A^{X_1}, B) \wedge \neg\mathbf{C}(A^{X_0}, B))$
$\quad \wedge (\mathbf{C}(A^{X_1}, B) \vee \neg\mathbf{C}(A^{X_1}, B))$

(9) $[(8), (p \vee q) \wedge (s \vee t) \equiv (p \wedge s) \vee (p \wedge t) \vee (q \wedge s) \vee (q \wedge t)]$

$\forall A \forall B \bullet \mathbf{C}(A^{X_0}, B) \wedge \mathbf{C}(A^{X_1}, B) \vee \neg\mathbf{C}(A^{X_1}, B) \wedge \mathbf{C}(A^{X_0}, B) \wedge \mathbf{C}(A^{X_1}, B)$
$\quad \vee \neg\mathbf{C}(A^{X_1}, B) \wedge \neg\mathbf{C}(A^{X_0}, B) \wedge \mathbf{C}(A^{X_1}, B)$
$\quad \vee \mathbf{C}(A^{X_0}, B) \wedge \neg\mathbf{C}(A^{X_1}, B) \vee \neg\mathbf{C}(A^{X_1}, B) \wedge \mathbf{C}(A^{X_0}, B) \wedge \neg\mathbf{C}(A^{X_1}, B)$
$\quad \vee \neg\mathbf{C}(A^{X_1}, B) \wedge \neg\mathbf{C}(A^{X_0}, B) \wedge \neg\mathbf{C}(A^{X_1}, B)$

(10) $[(9), (p \wedge \neg p) \equiv \texttt{false}; p \wedge p \equiv p]$

$\forall A \forall B \bullet \mathbf{C}(A^{X_0}, B) \wedge \mathbf{C}(A^{X_1}, B) \vee \texttt{false} \vee \texttt{false}$
$\quad \vee \mathbf{C}(A^{X_0}, B) \wedge \neg\mathbf{C}(A^{X_1}, B) \vee \neg\mathbf{C}(A^{X_1}, B) \wedge \mathbf{C}(A^{X_0}, B)$
$\quad \vee \neg\mathbf{C}(A^{X_1}, B) \wedge \neg\mathbf{C}(A^{X_0}, B)$

(11) $[(10), p \vee \texttt{false} \equiv p; p \vee p \equiv p]$

$\forall A \forall B \bullet \mathbf{C}(A^{X_0}, B) \wedge \mathbf{C}(A^{X_1}, B) \vee \mathbf{C}(A^{X_0}, B) \wedge \neg\mathbf{C}(A^{X_1}, B)$
$\quad \vee \neg\mathbf{C}(A^{X_1}, B) \wedge \neg\mathbf{C}(A^{X_0}, B)$

(12) $[(11),$ **Theorem 5.2.3**$]$

$\forall A \forall B \bullet \mathbf{C}(A, B^{X_0}) \wedge \mathbf{C}(A, B^{X_1}) \vee \mathbf{C}(A, B^{X_0}) \wedge \neg\mathbf{C}(A, B^{X_1})$
$\quad \vee \neg\mathbf{C}(A, B^{X_1}) \wedge \neg\mathbf{C}(A, B^{X_0})$

(13) $[(12),$ **Axiom 5.2.3**$]$

$\forall A \forall B \bullet \mathbf{C}(B^{X_0}, A) \wedge \mathbf{C}(B^{X_1}, A) \vee \mathbf{C}(B^{X_0}, A) \wedge \neg\mathbf{C}(B^{X_1}, A)$
$\quad \vee \neg\mathbf{C}(B^{X_1}, A) \wedge \neg\mathbf{C}(B^{X_0}, A)$

(14) $[(13), (\forall x \forall y \bullet p) \rightarrow (\forall y \forall x \bullet p)]$

$\forall B \forall A \bullet \mathbf{C}(B^{X_0}, A) \wedge \mathbf{C}(B^{X_1}, A) \vee \neg\mathbf{C}(B^{X_1}, A) \wedge \neg\mathbf{C}(B^{X_0}, A)$
$\quad \vee \neg\mathbf{C}(B^{X_1}, A) \wedge \mathbf{C}(B^{X_0}, A)$

(15) $[(14),$ **Definition A.0.1**, **A.0.2** *of* $<_s$ *and* $=_s]$

$X_0 =_s X_1 \vee X_1 <_s X_0$

(16) $[(1), (15)]$

$\forall X_0, X_1 \bullet \neg(X_0 <_s X_1) \rightarrow (X_0 =_s X_1 \vee X_1 <_s X_0)$
(17) $[(16), (p \rightarrow q) \equiv \neg p \vee q]$
$\forall X_0, X_1 \bullet X_0 <_s X_1 \vee X_0 =_s X_1 \vee X_1 <_s X_0$

Theorem 5.2.5. Let A and B be two object regions, **Front**(A,B), **Left**(A,B), **Right**(A,B) and **Behind**(A,B) are pairwise disjoint.

$$\mathbf{Front}(A,B) \wedge \mathbf{Left}(A,B) \equiv \mathtt{false}$$

Proof sketch:

(1) [**Definitions 5.2.6**, **5.2.7** *of* **Front** *and* **Left**]
$\mathbf{Front}(A,B) \wedge \mathbf{Left}(A,B)$
$\stackrel{\text{def}}{=} (\forall p | p \in \{B.\mathtt{left}, B.\mathtt{front}, B.\mathtt{right}, B.\mathtt{back}\}$
$\bullet p \neq B.\mathtt{front} \rightarrow \mathit{nearer}(A, B.\mathtt{front}, p))$
$\wedge(\forall p | p \in \{B.\mathtt{left}, B.\mathtt{front}, B.\mathtt{right}, B.\mathtt{back}\}$
$\bullet p \neq B.\mathtt{left} \rightarrow \mathit{nearer}(A, B.\mathtt{left}, p))$
(2) $[(1), \forall x \bullet p(x) \equiv p(x_0) \wedge p(x_1) \wedge \ldots]$
$B.\mathtt{left} \neq B.\mathtt{front} \rightarrow \mathit{nearer}(A, B.\mathtt{front}, B.\mathtt{left})$
$\wedge B.\mathtt{front} \neq B.\mathtt{front} \rightarrow \mathit{nearer}(A, B.\mathtt{front}, B.\mathtt{front})$
$\wedge B.\mathtt{right} \neq B.\mathtt{front} \rightarrow \mathit{nearer}(A, B.\mathtt{front}, B.\mathtt{right})$
$\wedge B.\mathtt{back} \neq B.\mathtt{front} \rightarrow \mathit{nearer}(A, B.\mathtt{front}, B.\mathtt{back})$
$\wedge B.\mathtt{left} \neq B.\mathtt{left} \rightarrow \mathit{nearer}(A, B.\mathtt{left}, B.\mathtt{left})$
$\wedge B.\mathtt{front} \neq B.\mathtt{left} \rightarrow \mathit{nearer}(A, B.\mathtt{left}, B.\mathtt{front})$
$\wedge B.\mathtt{right} \neq B.\mathtt{left} \rightarrow \mathit{nearer}(A, B.\mathtt{left}, B.\mathtt{right})$
$\wedge B.\mathtt{back} \neq B.\mathtt{left} \rightarrow \mathit{nearer}(A, B.\mathtt{left}, B.\mathtt{back})$
(3) $[(2), \mathtt{true} \rightarrow p \equiv p; \mathtt{false} \rightarrow p \equiv \mathtt{true}]$
$\mathit{nearer}(A, B.\mathtt{front}, B.\mathtt{left}) \wedge \mathit{nearer}(A, B.\mathtt{front}, B.\mathtt{right})$
$\wedge \mathit{nearer}(A, B.\mathtt{front}, B.\mathtt{back}) \wedge \mathit{nearer}(A, B.\mathtt{left}, B.\mathtt{front})$
$\wedge \mathit{nearer}(A, B.\mathtt{left}, B.\mathtt{right}) \wedge \mathit{nearer}(A, B.\mathtt{left}, B.\mathtt{back})$
(4) $[(3), p \wedge q \rightarrow p]$
$\mathit{nearer}(A, B.\mathtt{front}, B.\mathtt{left}) \wedge \mathit{nearer}(A, B.\mathtt{left}, B.\mathtt{front})$
(5) [(4), **Definition 5.2.4** *of nearer*]
$\exists X_1 \bullet \mathbf{C}(A^{X_1}, B.\mathtt{front}) \wedge \neg \mathbf{C}(A^{X_1}, B.\mathtt{left})$
$\wedge \exists X_2 \bullet \mathbf{C}(A^{X_2}, B.\mathtt{left}) \wedge \neg \mathbf{C}(A^{X_2}, B.\mathtt{front})$

(6) [(5), *Let* x_1 *be* X_1 *and* x_2 *be* X_2]
$\mathbf{C}(A^{x_1}, B.\mathtt{front}) \wedge \neg\mathbf{C}(A^{x_1}, B.\mathtt{left})$
$\wedge\mathbf{C}(A^{x_2}, B.\mathtt{left}) \wedge \neg\mathbf{C}(A^{x_2}, B.\mathtt{front})$
(7) [(6) $p \wedge q \equiv q \wedge p$]
$\mathbf{C}(A^{x_1}, B.\mathtt{front}) \wedge \neg\mathbf{C}(A^{x_2}, B.\mathtt{front})$
$\wedge\mathbf{C}(A^{x_2}, B.\mathtt{left}) \wedge \neg\mathbf{C}(A^{x_1}, B.\mathtt{left})$
(8) [(7), **Definition A.0.1** *of* $<_s$]
$x_1 <_s x_2 \wedge x_2 <_s x_1$
(9) [(8), **Theorem A.0.2**]
$x_1 <_s x_1$
(10) [(9), **Theorem A.0.3**]
`false`

The proofs of the existence and the uniqueness of *fiat* containers are similar to the proofs of those of A^X.

Index